# Springer Proceedings in Mathematics & Statistics

## Volume 64

This book series features volumes composed of select contributions from workshops and conferences in all areas of current research in mathematics and statistics, including OR and optimization. In addition to an overall evaluation of the interest, scientific quality, and timeliness of each proposal at the hands of the publisher, individual contributions are all refereed to the high quality standards of leading journals in the field. Thus, this series provides the research community with well-edited, authoritative reports on developments in the most exciting areas of mathematical and statistical research today.

For further volumes:
http://www.springer.com/series/10533

Jan Rychtář • Sat Gupta • Ratnasingham Shivaji
Maya Chhetri

Editors

# Topics from the 8th Annual UNCG Regional Mathematics and Statistics Conference

 Springer

*Editors*
Jan Rychtář
Department of Mathematics and Statistics
University of North Carolina
   at Greensboro
Greensboro, NC, USA

Sat Gupta
Department of Mathematics and Statistics
University of North Carolina
   at Greensboro
Greensboro, NC, USA

Ratnasingham Shivaji
Department of Mathematics and Statistics
University of North Carolina
   at Greensboro
Greensboro, NC, USA

Maya Chhetri
Department of Mathematics and Statistics
University of North Carolina
   at Greensboro
Greensboro, NC, USA

ISSN 2194-1009          ISSN 2194-1017 (electronic)
ISBN 978-1-4614-9331-0     ISBN 978-1-4614-9332-7 (eBook)
DOI 10.1007/978-1-4614-9332-7
Springer New York Heidelberg Dordrecht London

Library of Congress Control Number: 2013953295

Printed on acid-free paper

Springer is part of Springer Science+Business Media (www.springer.com)

# Preface

The Annual University of North Carolina Greensboro Regional Mathematics and Statistics Conference (UNCG RMSC) has provided a venue for student researchers to share their work since 2005. UNCG-RMSC is an annual one-day conference promoting student research in mathematics, statistics, and their applications in various fields. The 2012 conference was held on Saturday, November 3, 2012.

The conference was attended by a record number of 164 participants, of whom 78 were undergraduate students, 42 were graduate students, and 44 were faculty. The participants formed a very diverse pool: 73 were women, 25 were Asian, 20 were African American, and 2 were Hispanic. Participants came from 36 different universities and colleges. The schools with the biggest number of participants were UNCG (44), NC State University (15), Clemson University (12), Winthrop University (11), Bennett College (9), and Kennesaw State University (8).

The undergraduate students delivered a total of 30 presentations and the graduate students delivered 27 presentations. The talks were on various topics of mathematical biology, differential equations, statistics, biostatistics, number theory, algebra, combinatorics, applied mathematics, probability, and computational mathematics. The North Carolina Chapter of the American Statistical Association sponsored the best presentation competition. All presentations were evaluated by a group of faculty volunteers and the selected presentations are as follows:

| Undergraduate students: | Graduate students: |
| --- | --- |
| 1. Alison Miller, Elon University | 1. Virginia Burger, CMU-Pittsburgh University |
| 2. Chris Miles, Lafayette College | 2. John Steenbergen, Duke University |
| 3. Thomas Parrish, UNCG | 3. Andrew Snyder-Beattie, NC State University |

Apart from 57 student presentations, the conference featured three plenary presentations by invited faculty speakers:

- Katia Koelle, Duke University: The use of mathematical models to understand and control viral pathogens.

- Sujit Ghosh, NC State University: A Statistician's Journey Through the "Bayesian" Path.
- Michael Dorff, Brigham Young University: Kidney transplants, the Iron Man suit, and Pixar's movie "the Incredibles."

The conference would not have happened without the generous support of our sponsors. Funding and support for this conference were provided by the National Science Foundation (grant DMS–1229984), Mathematical Association of America (MAA) Regional Undergraduate Mathematics Conferences program (grant DMS–0846477), North Carolina Chapter of the American Statistical Association, Department of Mathematics and Statistics, UNCG, and the UNCG Office of Undergraduate Research.

All presenters were invited to submit a manuscript to this issue and the submitted papers subsequently went through a rigorous referee process. The topics covered in this issue mimic the main topics of the conference and the reader will find papers on differential equations, number theory, algebra, combinatorics, probability, statistics, mathematical biology, and computational mathematics.

The first four papers describe four different programs aimed at research with undergraduate students.

Dr. M. Dorff describes the highly successful national Center for Undergraduate Research in Mathematics (CURM) that was established in 2006 at Brigham Young University to support the undergraduate research nationwide. Dr. Sujit Ghosh describes the Computation for Undergraduates in Statistics Program at NC State University, Dr. Khan and his colleagues describe the Creative Inquiry established at Clemson University, and Dr. Crowe and her colleagues describe the math biology research program for undergraduate at UNCG.

The remaining papers have all substantial student coauthorship. Both the students and the mentors deserve a large applaud for the work they have done. It is not difficult to look beyond the papers to see the dedicated work of many faculty mentors who go well beyond their duties to attract students to research projects in mathematics and statistics. The mentors are now rewarded by the endless effort by those excellent students who completed their research projects and finished it by submitting and publishing their paper. Congratulations to all for this achievement!

Greensboro, NC, USA

Jan Rychtář
Sat Gupta
Ratnasingham Shivaji
Maya Chhetri

# Contents

1  CURM: Promoting Undergraduate Research in Mathematics .......  1
   Michael Dorff

2  NCSU-CUSP: A Program Making a Difference
   in Quantitative Sciences ...................................................  7
   Sujit K. Ghosh

3  Quantitative Methods in Biomedical Applications:
   Creative Inquiry and Digital-Learning Environments
   to Engage and Mentor STEM Students in Mathematics
   (NSF Funded Research) ...................................................  15
   Taufiquar Khan, John Desjardins, Marylin Reba,
   Ellen Breazel, and Irina Viktorova

4  Proving the "Proof": Interdisciplinary Undergraduate
   Research Positively Impacts Students ...................................  25
   M.L. Crowe, J. Rychtář, O. Rueppell, M. Chhetri,
   D.L. Remington, and S.N. Gupta

5  Modeling Heat Explosion for a Viscoelastic Material .................  31
   Irina Viktorova, Kyle Fairchild, and Jeff Fischer

6  Soliton Solutions of a Variation of the Nonlinear
   Schrödinger Equation .....................................................  39
   Erin Middlemas and Jeff Knisley

7  Galois Groups of 2-Adic Fields of Degree 12
   with Automorphism Group of Order 6 and 12 .........................  55
   Chad Awtrey and Christopher R. Shill

8  Laplace Equations for Real Semisimple Associative
   Algebras of Dimension 2, 3 or 4. ......................................  67
   James S. Cook, W. Spencer Leslie, Minh L. Nguyen,
   and Bailu Zhang

**9    Fibonacci and Lucas Identities via Graphs** ............................ 85
      Joe DeMaio and John Jacobson

**10   More Zeros of the Derivatives of the Riemann Zeta
      Function on the Left Half Plane** ......................................... 93
      Ricky Farr and Sebastian Pauli

**11   Application of Object Tracking in Video Recordings to the
      Observation of Mice in the Wild** ........................................ 105
      Matina Kalcounis-Rueppell, Thomas Parrish,
      and Sebastian Pauli

**12   The Card Collector Problem** ............................................. 117
      Anda Gadidov and Michael Thomas

**13   The Effect of Information on Payoff in Kleptoparasitic
      Interactions** .............................................................. 125
      Mark Broom, Jan Rychtář, and David G. Sykes

**14   A Field Test of Optional Unrelated Question Randomized
      Response Models: Estimates of Risky Sexual Behaviors** ............... 135
      Tracy Spears Gill, Anna Tuck, Sat Gupta, Mary Crowe,
      and Jennifer Figueroa

**15   A Spatially Organized Population Model to Study
      the Evolution of Cooperation in Species with Discrete
      Life-History Stages** ...................................................... 147
      Caitlin Ross, Olav Rueppell, and Jan Rychtář

**16   Analysis of Datasets for Network Traffic Classification** ............... 155
      Sweta Keshapagu and Shan Suthaharan

**About the Editors** ............................................................. 169

# Chapter 1
# CURM: Promoting Undergraduate Research in Mathematics

Michael Dorff

## 1.1 Introduction

In order to help more students and professors have a successful experience in doing undergraduate research in mathematics, the national Center for Undergraduate Research in Mathematics (CURM) was established in 2006 with NSF grants totaling over $2.5 million. CURM promotes academic year undergraduate research in the mathematical sciences by:

- training faculty members as mentors for undergraduate research projects;
- having these faculty members mentor undergraduate students in research groups that consist of two to five students who work together as a team on a research project during the academic year at their own institution; and,
- preparing undergraduate students to succeed in graduate studies in mathematics.

To help achieve this, CURM administers mini-grants annually to 15 professors working with about 45 undergraduate students on research during the academic year at various institutions across the USA. These mini-grants include a $3,000 stipend for each participating undergraduate student, a $6,000 stipend for each professor to reduce her/his teaching load in order to adequately mentor the group of students in research, and $250 in supply funds for each research group. Also, there is an annual summer workshop to train the professors in mentoring skills, and there is a culminating spring research conference in which the undergraduate students present their research, learn more about mathematics and opportunities available to those who study mathematics, and information to prepare them to attend and succeed in graduate school.

M. Dorff (✉)
Department of Mathematics, Brigham Young University, Provo, UT 84602, USA
e-mail: mdorff@math.byu.edu

J. Rychtář et al. (eds.), *Topics from the 8th Annual UNCG Regional Mathematics and Statistics Conference*, Springer Proceedings in Mathematics & Statistics 64, DOI 10.1007/978-1-4614-9332-7_1, © Springer Science+Business Media New York 2013

For the past 4 years, CURM has awarded mini-grants to 64 professors (41% female, 19% minority) from 54 different institutions to work with 195 undergraduate students (54% female, 29% minority). During the first 3 years of CURM (the years that have been completed), the 147 undergraduate students have written 60 joint research papers, 15 of which have been published in research journals such as Discrete Math, Journal of Difference Equations, Journal of Pure and Applied Math, International Journal of Biomathematics, Applied Probability Trust, and Involve while some of the other papers are currently being refereed. In addition, CURM students have given 123 single or joint conference presentations, 35 poster presentations, and have received 29 awards for their presentations or research. Finally, the data indicates that at institutions participating in the CURM program, about 18% of the math majors go on to graduate school while 63% of the CURM students at these schools go on to graduate school. More information about CURM can be found at its web site http://curm.byu.edu.

## 1.2   Center for Undergraduate Research in Mathematics

The benefits for students who participate in undergraduate research in a STEM field are significant as reports have shown [4,5,10,11]. These benefits can be summarized to include gains in knowledge and skills, academic achievement and educational attainment, professional growth and advancement, and personal growth [9]. For students from underrepresented groups, a research experience with an experienced faculty mentor is positively correlated with improvements in students' grades, retention rates, and motivation to pursue and succeed in graduate school [3, 6, 8]. Generally, there are two types of undergraduate research projects in mathematics: multiple-week summer REUs and individualized academic-year projects at the student's own institution. CURM offers another model.

### 1.2.1   Mini-Grants

CURM offers 15 mini-grants each year to faculty mentors who are accepted into the program. These mini-grants consist of training and financial support for undergraduate research groups consisting of two to five undergraduate students. These groups start during the fall semester and continue through the academic year. Typically, the students commit to work 10 h/week at their own institution on the research project for two semesters. The entire group meets at least 1 h a week and the students meet and work together at least 3 h a week. The rest of the time each individual student works on his/her research problem. CURM offers a $3,000 stipend for each student in the group ($1,000 to be paid at the beginning of the fall semester, $1,000 to be paid at the beginning of the spring semester, $500 to be paid after the student presents at the spring research conference, and $500 to be paid after submitting the final research paper/report). By having students work together in groups, they tend to motivate each other and they also learn to become more independent of the faculty mentor. Of course, the faculty mentor needs to be actively involved with the group. However, many of them are at institutions with a teaching load of three to four courses per semester. Hence, CURM provides $6,000 for the professor to buy out at least one course from his/her teaching load during the academic year in order to free up time to spend working with these mentored groups.

## 1.2.2   Summer Training Workshop

Before the faculty members begin mentoring their students in the undergraduate research group, they attend a 2-day summer workshop. The purposes of this workshop are to discuss effective approaches in working with undergraduate students in academic year research and develop a rapport among the professors. There are specific presentations and discussions lead by the CURM directors. These include such topics as developing a manageable timeline for academic year undergraduate research, how to get started mentoring undergraduate students in research, potential pitfalls and overcoming them during the mentoring journey, working with group dynamics among students with different backgrounds and skills, and helping students develop independence in doing research. Some of this has been published in papers co-authored by various CURM directors and CURM professors [1, 2, 7]. Finally, there is a CURM Facebook page that the professors are invited to join to facilitate discussions about their experiences.

## 1.2.3   Spring Research Conference

Having the undergraduate students present their research in a supportive environment is very beneficial in motivating them to be consistent in their research, to feel the excitement of the mathematical community, and to prepare for graduate school. Therefore, we organize a CURM research conference that each student participant and faculty mentor attends. The conference consists of activities to motivate and intellectually stimulate students to continue to study mathematics and prepare for graduate school, and 20-min sessions in which the student participants present their research with written feedback and guidance from two CURM professors.

In the past, this conference has been held at Brigham Young University (BYU) in March. We have brought in three keynote speakers known for giving interesting mathematical talks appropriate for undergraduate students, such as Bob Devaney, Joe Gallian, Aparna Higgins, Colin Adams, Laura Taalman, Tony DeRose, Dave Kung, and Frank Morgan. Also, we have created the *What is ... ?* series in which professors, who are known for being excellent teachers and being able to connect with undergraduate students, give a 30-min presentation on advanced topics at a level suitable for the students. Topics have included hyperbolic geometry, operations research, minimal surfaces, coding theory, cryptography, and dynamical systems. There have been panel discussions on attending graduate school in mathematics with panelists. To help students build a sense of belonging to the mathematics community, there were also social activities such as a banquet, a hike, and a reception with games.

### 1.2.4   Research Reports

Having the undergraduate students write up a paper about their research is very beneficial both in motivating them to be consistent in their research and in preparing for graduate school. In the research paper, we encourage the group to not only describe their research but also propose some open problems in the research area that they would have worked on if they had more time. Thus, the paper is not only a tangible end product for the initial research group but is also a written starting resource with a set of research problems for future undergraduate students who are recruited to work with that professor on research. We require all groups to submit to CURM a final written research paper at the end of May of the academic year. We encourage the CURM professors to have their undergraduate students submit their research papers to refereed journals for publication if appropriate.

## 1.3   The Effectiveness of the CURM Program

CURM has a tremendous effect upon undergraduate students, their professors, and their departments and institutions. It has been fantastic to witness the indirect benefit of changing the practices and culture in mathematics departments and in some cases even at institutions as groups have participated in the CURM program. As CURM students and professors have shared the results and experiences in doing undergraduate research (e.g., presentations of their research in the department, awards from their institution or from conference presentations, university newspaper articles, acceptance to graduate school, etc.), other students have listened and have become interested in doing research. This has resulted in some departments creating new courses in which students now can get academic credit for doing research allowing a professor to count that research as part of his/her teaching assignment. In some places, the dean has been impressed with the results of the CURM research group and has offered internal funds to the CURM professor to continue to work with undergraduate students on research after the CURM year is over. At two minority-serving institutions, Jackson State University and California State University—Channel Islands (CSUCI), the administrations have been so impressed with the CURM program that they have introduced new university-wide programs promoting undergraduate research in all disciplines based upon the CURM model. Below are some remarks by previous CURM participants on the effect the CURM program has had.

## 1.3.1   Undergraduate Students

CURM has opened many doors for my future. It encouraged me to apply for a summer 2008 REU [got accepted into one, and attended it]... If it weren't for CURM, I wouldn't be where I am today; I wouldn't know what it meant to do research, and I wouldn't be applying for graduate school.
>   Amy Stockman, Concordia University

At my institution, students are mainly Hispanic origin. I have seen many of my students struggling at school because they had to work outside of the classes to support themselves or their families. This has tremendous impact on their academic achievements. Most of them even do not think about continuing on higher education ... CURM provided the hand I needed to extend to my two female students ... both of them will be the first generation who will be going to a graduate school among their family members.
>   Gulhan Alpargu, California State University–Fullerton

[I want to mention] how important CURM grant was for me and my students here. Couple of years ago there were [hardly any] students thinking of maybe applying to a graduate program in the future, but now we have at least 2–3 per year that are actually taking the GRE tests and applying for graduate schools.
>   Nicoleta Tarfulea, Purdue University–Calumet

## 1.3.2   Faculty and Institutions

Because of the CURM grant, I was able to work with a large number of students (7 total, while only 2 were supported by CURM). All it took was this one year of the CURM grant to fan the fire, and our department has begun to foster an environment that encourages undergraduate research. This coming year there will be 4 professors working with students or groups of students on research projects.
>   Joan Lind, Belmont University

After learning of my CURM group, the Dean of Faculty at CSUCI introduced a pilot undergraduate research program where faculty in any discipline may apply to receive teaching credit for offering a course where students work on research projects. More recently, the Dean has established a Student Research Steering Council to embed student research experiences across the curriculum. In other words, institutionalization of undergraduate research at my university has been sparked by my CURM experience.
>   Kathryn Leonard, California State University–Channel Islands

**Acknowledgements**   CURM has been funded by NSF grants DMS-0636648 and DMS-1148695 and by Brigham Young University.

# References

1. Bailey, B., Budden, M., Dorff, M., Ghosh-Dastidar, U.: Undergraduate research: how do we begin? MAA Focus **29**(1), 14–16 (2009)
2. Bailey, B., Budden, M., Ghosh-Dastidar, U.: Practical tips for managing challenging scenarios in undergraduate research. MAA online column Resources for Undergraduate Research, no. 3. Available at http://www.maa.org/external_archive/columns/Resources/resources_12_08. html (December 2008)
3. Barlow, A., Villarejo, M.: Making a difference for minorities: evaluation of an educational enrichment program. J. Res. Sci. Teach. **41**, 861–881 (2004)
4. Hathaway, R.S.: The relationship of undergraduate research participation to graduate and professional educational pursuit: an empirical study. J. Coll. Stud. Dev. **43**, 614–631 (2002)
5. Hunter, A.-B., Laursen, S.L., Seymour, E.: Becoming a scientist: the role of undergraduate research in students' cognitive, personal, and professional development. Sci. Educ. **91**, 36–74 (2006)
6. Ishiyama, J.T., Hopkins, V.M.: Assessing the impact of a graduate-school preparation program on first-generation, low-income college students at a public liberal arts university. J. Coll. Stud. Ret. **4**, 393–405 (2002)
7. Leonard, K.: Adventures in academic year undergraduate research. Not. Am. Math. Soc. **55**(11), 1422–1426 (2008)
8. Nagda, B., Gregerman, S., Jonides, J., von Hippel, W., Lerner, J.: Undergraduate student-faculty research partnerships affect student retentions. Rev. High. Educ. **22**, 55–72 (1998)
9. Osborn, J.M., Karukstis, K.K.: The benefits of undergraduate research, scholarship, and creative activity. M. Boyd and J. Wesemann (eds.) In: Broadening Participation in Undergraduate Research: Fostering Excellence and Enhancing the Impact, pp. 41–53. Council on Undergraduate Research, Washington, DC (2009)
10. Seymour, E., Hunter, A.-B., Laursen, S.L., DeAntoni, T.: Establishing the benefits of research experiences for undergraduates: first findings from a three-year study. Sci. Educ. **88**, 493–534 (2004)
11. Sharp, L., Kleiner, B., Frechtling, J.: A description and analysis of best practice findings of programs promoting participation of underrepresented undergraduate students in science, mathematics, engineering, and technology fields. Report No. NSF 01-31. NSF, Arlington (2000)

# Chapter 2
# NCSU-CUSP: A Program Making a Difference in Quantitative Sciences

Sujit K. Ghosh

*AMS Subject Classification:* 62F03, 62F15, and 62P10

## 2.1 Introduction

The Department of Statistics at North Carolina State University (NCSU) established a Computational Science Training for Undergraduates in the Mathematical Sciences (CSUMS) program funded by the National Science Foundation (NSF) under the leadership of the Principal Investigator, Professor Sujit K. Ghosh. The overarching goal is to provide a rich applied computational statistics research experience to a diverse population of undergraduate students that will encourage them to continue their academic programs to the graduate level and will help them in making more informed decisions about their academic or nonacademic careers.

The NSF-CSUMS project titled NCSU Computation for Undergraduates in Statistics Program (NCSU-CUSP), prepares students to engage in a significant research experience, and to be fluent in the languages of computing, mathematics, and statistics. The program was launched on September 15, 2007, with funding from the prestigious NSF-CSUMS award and the program has been awarded a total of $770,714 to date (Award# NSF-DMS 0703392: http://www.nsf.gov/awardsearch/showAward?AWD_ID=0703392). NCSU-CUSP targets rising senior and junior mathematics majors at NCSU and Meredith College who have demonstrated academic excellence. With rapid advances in technology, massive amounts of new data are generated daily in many scientific disciplines and the volumes are growing at a rate unprecedented in human history. For the USA to remain

S.K. Ghosh (✉)
Department of Statistics, North Carolina State University, Raleigh, NC 27695-8203, USA
e-mail: sujit_ghosh@ncsu.edu

J. Rychtář et al. (eds.), *Topics from the 8th Annual UNCG Regional Mathematics and Statistics Conference*, Springer Proceedings in Mathematics & Statistics 64, DOI 10.1007/978-1-4614-9332-7_2, © Springer Science+Business Media New York 2013

competitive and innovative, a diverse pool of researchers trained in novel and powerful techniques is critically needed to illustrate, model, and analyze these large-sized, high-dimensional, and nonlinearly structured data.

Building on resources of one of the country's largest statistics departments, NCSU-CUSP has become one of the first computationally intensive statistics programs for undergraduates in the nation. The cutting-edge projects from this program have led to the development of new computationally intensive courses and interdisciplinary courses, which will have a long-term impact. The project is also committed at the outset to increasing diversity in the emerging field of computational statistics. NCSU-CUSP has increased awareness of statistical science among mathematics majors and faculty, it has fostered greater collaboration between interdisciplinary programs, and it has encouraged a diverse pool of well-prepared students to pursue graduate studies in quantitative sciences. To date, the program has supported 34 undergraduate students who worked in a cohort of 6–8 students in each academic year since summer of 2008. Out of these 34 students, 27 ($\sim$80 %) are female out of which 2 are African-American students. Out of the 28 students who have completed the program, all of the graduating seniors have either entered into a Masters' program or a PhD program in Statistics, and a majority of them have chosen NCSU as their graduate program.

The project has supported four bright students and a faculty member from the local Meredith College, which has become one of the largest independent private women's colleges in the United States of America (USA). A letter from Dr. E. Jacquelin Dietz (Professor and Head of the Department of Mathematics and Computer Science at Meredith College) describes the impressive impact and contribution of the NCSU-CUSP. In particular, Professor Dietz remarked "The rich experiences that (NCSU)CUSP provided them (Meredith students) in statistics, mathematics, computation and genetics will inform and inspire their teaching of future generations of young students." The NCSU-CUSP has also supported faculty members with partial salaries (during summer) who have served as the lead instructors and mentors over the past 4 years. The instructors have expressed a strong sense of satisfaction and motivation to work with the young students.

The NCSU-CUSP begins with a 10 week summer program that usually starts from the end of May through the end of July each year. During this period, three to four teams of two to three students work collaboratively with program faculty mentor. Topics explored to date range from environmental statistics (e.g., "Investigation of blood lead levels in children") to financial statistics (e.g., "Dynamics of credit ratings") to statistical genetics (e.g., "Optimization of Grammatical Evolution Decision Trees for detecting Epistasis" and "Comparison of analytical methods for genomic association studies"). The program couples extensive coursework throughout the academic year in computing for contemporary statistical analysis with a practicum and research lab focusing on an area of application mentioned above. Dr. Alison Motsinger-Reif has been leading the program on Statistical genetics projects for the past 2 years and in a supporting letter she succinctly summarized the broad impact of the program in making a tremendous difference with Statistics department and beyond. In particular Dr. Motsinger-Reif commented

"The successes of (NCSU)CUSP are clear, in many ways. The program has supported some of the most talented undergraduates at the university in performing high-quality research.... This (financial support) ensures gifted students from many economic background are able to participate in the program."

The NCSU-CUSP not only provided financial support to all of the enrolled students but also helped them to develop skills in data management and manipulation, converting data to a form convenient for statistical analysis, and to develop simple to complex statistical procedures and graphics. Training in communication skills helped to develop graduates who can bring scientific research results to the public and policy makers. The students benefitted from a significant, collaborative interdisciplinary scientific research experience under the mentorship of faculty working at the forefronts of their disciples. NCSU-CUSP supported all student travel allowing them to present their research work at regional, national, and international conferences. The activities of NCSU-CUSP are consistent with the recommendations of the important National Academy of Sciences' publication "Rising Above The Gathering Storm: Engaging and Empowering America for a Brighter Economic Future." Through this program, it has developed one of the first undergraduate-level computationally intensive and research oriented statistics curricula in the nation. In summary, the NCSU-CUSP has made a significant difference by

1. preparing undergraduate statistics/mathematics majors (in particular by engaging and encouraging women in mathematical fields) to take advantage of computing advances and make sophisticated computing an integral part of the curriculum and a significant research experience;
2. improving students' nontechnical skills, including public speaking, written communication, ethical reasoning, and the ability to creativity in developing statistical and computing approaches to solving interdisciplinary scientific research problems and
3. preparing and motivating a diverse pool of highly qualified students to pursue interdisciplinary graduate studies in the quantitative sciences.

## 2.2  Program Activities and Findings

There are several key aspects of the program that have lead to its successes. The program is very vertically integrated, with the program PI allowing the instructors freedom to run their cohorts in a way that best fits the subdiscipline that they are focusing on. The instructors work together as a team (there is a lecturer and a computing instructor) to teach students the skills they need for their research projects. Additionally, there have been graduate teaching assistants that have volunteered to help in the mentoring process by working with research teams. The undergraduates get the advantage of the expertise of all the team around them, and the graduate students get valuable experience in mentoring a research project. This integration also really educates the students about the process of continuing on in academia. By working so closely with both faculty and graduate students for a full year, students get lots of opportunities to learn about life and expectation at each level. Demystifying the field helps them make a more informed decision about moving through the academic pipeline to the next step. The structure and length of the program also helps to keep the students stay in the pipeline. NCSU-CUSP is not just a summer program, its a year-long so that students have help/guidance in applying for graduate school (letters, mentoring, etc.) in their senior year.

New courses have been developed in response to CSUMS activity and are open to all students. In particular, the statistical computing and data management course (which has become a required course for CSUMS students) provides a solid background on the use of computers to manage, process, and analyze data. The courses developed as a part of CSUMS activity are popular on campus and provide a strong foundation in statistics and computing needed to implement computationally intensive statistical methods. These courses have broad impacts on student training as they motivate a diverse pool of highly qualified students to pursue interdisciplinary graduate studies in the quantitative sciences. In particular, the environmental statistics practicum course has motivated students to pursue research activities in collaborations with scientists at US Environmental Protection Agency (EPA). Also the statistics and financial risk practicum course has been instrumental in motivating the students to understand and explore the mechanics of financial risk. Also, during the summer of 2010, by taking the course "Statistical Genetics Practicum," the students learned about computer-intensive data-mining tools for gene-mapping in human genetics and explore the relative performance of these methods on both real and simulated data.

The entire group of six to eight students met with their faculty advisers on a regular basis to discuss the project updates. All of them worked together on research projects, shared their research findings, collaborated on performing statistical simulations, and explored the impact of scientific theory using computational approaches. The insights gained in this collaborative exploration involving all students were then used to construct various statistical models. Computational tools included statistical modeling and data-mining software. The graduate student assigned to this project helped all eight students to learn computational methods executed via SAS. In addition, they also explored the consequences of the choice of distribution on the value of commonly used statistical metrics. All of the eight students have not only used conventional techniques as far as possible but they also used computer simulations to answer questions that these techniques cannot answer. Students were introduced to computational tools using local computing, as well as high-performance computing using NCSU's supercomputing cluster. Additionally, all students attended field trips to see how various data are collected, as well as to super-computing facilities to see how high performing computing is made possible. Although all eight students actively collaborate with their faculty mentors on all three projects, a smaller subset of students work in teams of two on the research projects. Later a smaller subset of students took the lead in writing the manuscripts for possible publication in undergraduate research journals.

The Statistics department at NCSU continues the development of its web site to publicize the department's CSUMS program: http://www.stat.ncsu.edu/cusp/. The web site is intended to blend with other departmental web sites devoted broadly to the academic, research, and human resource aspects of the department. The web site provides a detailed list of CSUMS activities including program objectives, information on financial aid, courses, research projects, and a photo gallery.

The undergraduate program has a Stat Club that serves both pre-professional and social/group cohesiveness functions. One of these meetings included a presentation focusing on graduate education, its benefits in expanded career opportunities, suggestions for preparing one's self for graduate study, and information about assistantship, fellowships and traineeship and their associated benefits. Other meetings featured speakers, sometimes past graduates of the program, who discuss their experiences as statisticians, what aspects of their training were especially valuable, and the opportunities they see for future graduates of our program. The Stat Club took a trip to the Washington DC area to visit Federal agencies that employ statisticians. The group also met with the board of directors of the American Statistical Association where they discussed the opportunity for graduate study in statistics. The undergraduates were chaperoned on this trip by a post doctoral fellow.

## 2.3   Program Impact

The program has encouraged participants to continue to graduate school, with ALL of the graduates of the program continuing to a graduate program in a quantitative area. This program has also had a direct impact on the career goals of several students who would not have attended graduate school. The program has also had an impact beyond the participants and helped expand NCSU's course offerings in key areas. The CUSP program recruits six of these students from NCSU each year. The impact on these students is clear. ALL of the students who participate in CUSP go on to graduate programs in quantitative fields. The extensive undergraduate research training along with training in advanced computing makes CUSP students attractive candidates for masters and PhD programs. In the first two cohorts all of the participants went on to graduate programs. In later cohorts some students have not yet graduated but are planning to go on to graduate programs. Several of the students went on to the NCSU graduate program in statistics. These students have exceptional academic records and compete well among the other students in our highly competitive program. A large number of the students went on to become part of the Masters of Science in Analytics (MSA) program at NCSU. The director of this program has expressed how the computational training and understanding of how to apply statistical theory to poorly defined problems has made our students exceptional members of the MSA program. Other students went on to graduate programs around the country. Almost every CUSP student was accepted at multiple graduate institutions. CUSP achieves this impact by bringing the participants together as a cohort that works on research projects. This cohort mentality allows the students to see role models not only in the faculty with whom they are working but also among their peers. That allows them to envision themselves working on research in graduate school that previously seemed daunting.

Overall, the CUSP program helps participants realize their desire to continue into graduate school. Obviously there may be some selection bias in this result in that many of the students who are involved in CUSP are very strong students who may have gone on to graduate school in quantitative areas regardless of their participation. However, in my role as academic advisor I have seen several students who changed their long-term plans because of the CUSP programs. For example, prior to participating in CUSP a student from the 2008 to 2009 cohort had discussed her long-term career plans with me as part of our normal advising meetings. At that time she felt that she was unsure as to how she would use statistics in a career. Although she was doing well in the theoretical courses in the undergraduate program she just did not see how these courses would apply to a "real-world" problem. This sentiment is common among our students who tend to come into statistics with a desire to solve real-world problems using quantitative methods. At that time, the student felt she would probably not go on to graduate school but instead seek employment. However, the CUSP program exposed her to how the more theoretical aspects of statistics can be translated into advanced problems in statistical methods. The program also exposed her to mentors who were working with advanced methods to solve "real-world" problems. This exposure reinvigorated her interest in academic pursuits. The student went on to the MSA program and now has started on a distinguished career as an Analytical Engineer at The SAS Institute. This impact of the CUSP program is not unique to a student. We are convinced that at least one other student in each of the cohorts would not have gone on to graduate programs if they had not participated in this program.

CUSP has also had an impact outside of the students who are directly involved in the program. Traditionally, we had offered a single introductory course on statistical computing. Over the last few years many students have expressed interest in taking more courses in statistical computing. As part of the CUSP program the department now offers a second course that gives students training in more advanced statistical computing methods. This course is extremely popular with many of our undergraduate students outside the CUSP program. Through this course, the CUSP program has had an impact that is much broader than the six students enrolled in the program each year.

## 2.4 Conclusions and Discussions

Aided by rapid advances in technology, massive amounts of new data are being generated daily across multiple scientific disciplines and are growing at an exponential rate unprecedented in human history. Researchers trained in novel and powerful techniques are critically needed to illustrate, model, and analyze these large-sized, high-dimensional, and nonlinear-structured data. NCSU-CUSP has increased the awareness of statistical science among minority mathematics majors and faculty, fostered greater collaboration between departments, and encouraged a diverse pool of well-prepared students to pursue graduate studies in quantitative sciences.

CUSP is a model for how programs to improved undergraduate research should work. Many programs implement Research Experiences for Undergraduates (REUs). CUSP goes beyond this typical model by incorporating a cohort structure that provides a built-in support mechanism for participants. It also supplements traditional research activities with new courses that train students in methods that they can apply immediately. This cohort structure combined with curricular transformation creates a model that can make undergraduate research work elsewhere. We firmly believe that CUSP is a program that is making a real difference among the students at NCSU and can serve as a model for real transformation at other institutions.

**Acknowledgments** This material is based upon the work supported by the National Science Foundation under grant number DMS-0703392. Any opinions, findings, and conclusions or recommendations expressed in this material are those of the author and do not necessarily reflect the views of the National Science Foundation.

# Chapter 3
# Quantitative Methods in Biomedical Applications: Creative Inquiry and Digital-Learning Environments to Engage and Mentor STEM Students in Mathematics (NSF Funded Research)

Taufiquar Khan, John Desjardins, Marylin Reba, Ellen Breazel, and Irina Viktorova

## 3.1 Introduction

Research in science and engineering is increasingly reliant on mathematical and statistical tools. The NSF has argued that to build a competitive international workforce in STEM fields, colleges and universities must inspire a greater number of students to learn a greater amount of mathematics and statistics [1]. The growing field of biomedical science and bioengineering challenges students to make critical decisions about people's lives and diseases and demands a deep understanding of the quantitative complexity both of the biological system and of the decision-making process. Biomedical science and bioengineering as well as other medical majors are among the most popular fields for college graduates today. For students to succeed in such fields, mathematicians must do a better job of explaining to students how mathematical concepts and quantitative analysis can be applied in biomedicine and why it is important to succeed in the undergraduate mathematics curriculum. The challenge is to catch the attention of STEM students by offering early applied learning experiences that engage them with the application of mathematics and statistics in professional practice and applied learning applications.

At a freshman or sophomore level, it can be a challenge to connect mathematical concepts with bioengineering and medical applications, and to challenge the students' view of what mathematics can offer them. Many incoming freshmen declare a STEM major, but know little about their declared field or about how Calculus can be applied to a particular STEM field. Students can be insufficiently motivated to work consistently in their Calculus courses in pursuit of undefined educational or life-long goals. Consequently, they can underachieve in these fundamental STEM

T. Khan • J. Desjardins • M. Reba • E. Breazel • I. Viktorova (✉)
Department of Mathematical Sciences, Clemson University, Clemson, SC 29634, USA
e-mail: iviktor@clemson.edu

J. Rychtář et al. (eds.), *Topics from the 8th Annual UNCG Regional Mathematics and Statistics Conference*, Springer Proceedings in Mathematics & Statistics 64, DOI 10.1007/978-1-4614-9332-7_3, © Springer Science+Business Media New York 2013

courses and possibly leave their STEM field. Too often, this STEM-attrition scenario disproportionally involves women, undeserved minorities, first-generation college students, and community-college transfer students [2]. The authors believe that one benefit of using medical applications in applied learning environments is their appeal to a broad range of students, as most everyone has personal experiences with health issues.

Past research has focused on the importance of success in the first college math course and its correlation with success in engineering, and other STEM fields [3]. Calculus is particularly noted to be a stumbling block [4]. Since 2006, the efforts by the authors' home institution have been heavily invested in classroom redesign of freshman Calculus courses. All sections of Calculus I adopted a variation of the SCALE-UP active-learning instructional model which includes short lectures, student collaboration at round tables, and graded group activities [5]. These changes are consistent with research emerging from the Calculus Reform Movement showing that the longer you lecture the less students retain, as well as with recommendations to include small-group or collaborative classroom learning activities [6]. Initial results with this revised curriculum have been very promising, with 2008 results showing a nearly 50 % reduction in the DFW (students receiving a D, F, or withdrawing from the course) rate compared with Fall 2005 measures. Despite these efforts however, approximately 20 % of students continue to earn a DFW, and had to either repeat the course or abandon their STEM career goals. Clearly, more innovative concepts in instruction should be considered to decrease this rate of student loss.

Recent work in the authors' home department has included the introduction of Tablet PCs into several sections of Freshman Math courses in 2006. Student perceptions, behaviour, and performance (especially of weaker students) were shown to improve [7]. With this the department created a dedicated technology classroom that included workstations with high-powered software, multiple projection capability, Smart screens, as well as Tablet PCs.

Another challenge to innovative and supplemental instruction is developing a learning opportunity that can fit it into a student's schedule and course-credit structure, and to insure that all participants (faculty and students both) receive merit-based credit for participation. At the authors' home institution, "Creative Inquiry" is a program course structure, which strives to engage students in the process of learning and discovering through faculty-mentored research and outreach activities across multi-disciplinary departments. Students who participate in these Creative Inquiry classes have been shown to learn and to think in new ways, learn non-class skills designed toward their interests, enhance their academic performance in other classes, improve their satisfaction with their learning environment, and improve their relationships with faculty. In addition, instructors who teach Creative Inquiry classes develop mentoring relationships with students, have the opportunity to develop courses toward a specific area of interest that spans several departments, and rejuvenate and improve their teaching in other courses. At the authors' home institution, the university provides monetary support for courses taught under the Creative Inquiry framework, and since its conception in 2005 the university has offered a total of 275 Creative Inquiry courses 12 of which are from the Mathematical Sciences department.

In this paper, the authors describe the 2-year NSF funded collaborative project between faculty from the Mathematical Sciences and Bioengineering departments that combines inspiration in Biomedicine with retention in Calculus, directed at freshman and sophomore students. This paper describes the initial results from Module 1: Orthopedics and Pre-Calculus and Module 3: Health Hazards from Arc-flash.

## 3.2  Methods for Module Organization

### 3.2.1  Program Structure

Students participated in 1-h modules where they discuss biomedical applications to their current math courses interact with faculty and student mentors, participate in field trips, and have access to a textbook repository. The goal of this program is to have all participants engaged in the interplay of mathematical and biomedical concepts in the context of interesting applications that may help them formulate career goals while deepening their understanding. This program was designed to emphasize mathematics and statistics relevant in four biomedical areas that are directly linked to the students progression through their core calculus courses:

Module 1:   Orthopaedics
Module 2:   Disease epidemiology
Module 3:   Health hazards from arc-flash
Module 4:   Mammography and radiology

Students have the opportunity to enroll in one module per semester for up to four semesters. They enter modules coordinated with their current or previous math courses [whether pre-calculus, first semester calculus (calculus of one variable), second semester calculus (calculus of one variable II), or third semester calculus (calculus of several variables)]. By presenting interesting biomedical problems as early undergraduate applied learning experiences, instructors are required to decompose a difficult mathematical problem into its simpler parts that students can manipulate. These modules are broken down into 5–7 weekly lessons of 1–2 h each. The modules usually begin 3–5 weeks into the semester, to give students an opportunity to learn the basics in their math courses before beginning these applied-learning experiences. Students are introduced to the exciting field of study and given an interesting problem to solve, with the mathematical component structured to their level of understanding. Students work through a problem, identify what they don't fully understand and seek remedies. They are then given the opportunity to interact with their peers in group activities and their instructors and participate in on- and off-campus field trips. This level of communication, where students work on a problem of interest, invest in learning, and even discuss future learning for their problem of interest, is impossible to achieve in large content heavy math classrooms that have little time to spare.

In addition to these group activities, students have the opportunity to visit local professional facilities that provide the services studied in the module. These field trips are undertaken as the students have the opportunity to explore the mathematical concepts that are relevant to the applied learning experience of interest and are used to reinforce the practical applications and empowering nature of the mathematical skills that they are acquiring in class and applying in their learning modules.

All participants receive mentoring from their active learning experience instructor and a designated advanced undergraduate mentor who works to enhance their success in their mathematics courses. All participants are matched with faculty or undergraduate-student mentors who will communicate with them both in-person and through web-based technology at various times throughout each week. Although the modules do not begin until 3–5 weeks into the semester (other than an introductory meeting on week 1), the mentoring begins the first week of classes.

Copies of textbooks from the core calculus courses are also available for student loan. Participation in the modules allows students to use these textbooks for the semester the module is taken. The library consists of texts from pre-calculus, single-variable calculus, multi-variable calculus, and topic specific biomedical and statistics texts that enhance their individual learning module experience.

### 3.2.2  Module 1 Curriculum

#### 3.2.2.1  Orthopedics: Fundamentals of Pre-Calculus in Orthopaedic Medicine

This beginning module was intended to reinforce pre-calculus curriculum based on the home institutions state pre-calculus standards. It was offered as a one credit course that was spread out over one semester and it was intended to engage the student in basic bioengineering problems requiring algebra and trigonometry, and introduce areas of study and applied mathematics that required the use of pre-calculus to effectively solve real-world problems. It reinforced scalar, algebraic and trigonometric concepts that were relevant to orthopaedics and total joint replacement. Students participating in this module were expected to have already taken or current be taking course content equivalent to the following university level courses: MTHSC 103 Elementary Functions, MTHSC 104 College Algebra, and MTHSC 105 Precalculus.

### 3.2.3   Module 1 Course Schedule

Week 1: Orientation and Introduction to Module
   (1 h with introduction, orientation and syllabus)
Week 3: Tour of Clemson Bioengineering Department and Biomechanics Lab
   (pre-survey and department tour, 1 h)
Week 4: Activity 1: Orthopaedics, Angles and Basic Trigonometry
   (15-min lecture with 45-min applied learning activity)
Week 5: Tour of Local Orthopaedics and Sports Medicine Practice (2 h)
Week 6: Activity 2: Anthropometry, Measurement, Percentiles and Averages
   (15-min lecture with 45-min applied learning activity)
Week 7: Student K-12 Outreach Project Development (1 h)
Week 8: Student K-12 Outreach Project Development (1 h)
Week 9: Activity 3: Orthopaedics, Angles and Polynomials
   (15-min lecture with 45-min applied learning activity)
Week 10: Tour of Total Joint Replacement Testing Facility (1 h)
Week 11: Student K-12 Outreach Project Development (1 h)
Week 12: Total Joint Replacement Motion and Kinematics
   (15-min lecture with 45-min applied learning activity)
Week 15: Student K-12 Outreach Presentations (one presentation and review)
Week 16: Module Review and Assessment (1 h summary and assessment)

### 3.2.4   Module 1 Activity Example and Details

#### 3.2.4.1   Week 4 (1 H Applied Learning Module: Orthopaedics, Angles and Basic Trigonometry)

In this module, students were given an opportunity to participate in a "life or death" project that challenged them to formulate a treatment regimen for an orthopaedic condition. This condition, known as a lower limb deformity, required the student to apply simple concepts in angle measurement and trigonometry to correct a bony anatomical deformity in a patient. This module began with in-class review of basic geometry and trigonometry and an introduction to the pathologic conditions of lower limb deformity. Students were then given a "patient's" X-rays that showed a common deformity of the lower limb. They were then asked to calculate a tibial re-alignment treatment to correct the deformity. Using these X-rays, the students used simple measures of bone length, width, and angular deformity, apply basic trigonometry to "cure" the patient. The accuracy of the surgical correction was then visualized on a surgical training bones and a computer model of this bony system. The students were encouraged to explore a range of treatment options using these interactive models. Discussions of actual before-after surgical treatments for this condition using X-rays were presented. These concepts contained some of the challenges for pre-calc students and therefore the reiteration and application of these topics was intended to help strengthen their understanding.

#### 3.2.4.2   Week 4 Outcomes

Hands-on use of rulers and protractors. Applied knowledge of scalar quantities, radians and degrees, relative and absolute angles, applied use of sines, cosines, tangent functions. Participation in Team Activity. Discussions of experimental variables. Mathematics applied: basic trigonometric functions

### 3.2.5   Student Projects

Students worked in-class and out-of-class in groups of 2–4 on class presentations which demonstrated their understanding of the various heat propagation models discussed in class, but did so in a pedagogical context where they tried to find creative ways of explaining this material and the underlying formulas to college students just beginning to learn calculus. Student and Instructor Reviews of all projects were tabulated; instructors announced which student ideas would be incorporated in the future development of the module, presentation at the conference or the journal publication [8].

## 3.3   Assessment

Assessment focused on how this Creative-Inquiry project-based approach, combined with introducing the students to mathematical skills they will need to learn (in some cases next semester), enabled the student to more confidently approach an entire mathematical concept in the context of applied learning.

Formative evaluations began with the first teaching of Modules and will continue with every implementation of each module. External evaluations are to take place midway through the 2-year program and again at the end of the program. Both evaluations are designed to gather information in order to answer the following questions:

- Goal 1: Does participation in these activity-based learning modules improve student knowledge in current math courses?
- Goal 2: Do these modules improve student performance in current math courses?
- Goal 3: Does participation in these activity-based learning modules improve student performance in future math courses?
- Goal 4: Does the implementation of activity-based learning using medical applications affect the retention in STEM majors?
- Goal 5: Do applied learning modules, such as the ones proposed, have disseminative potential to high-school, community college and other 4-year institutions with an interest in adopting this approach to enhance early undergraduate applied learning?

### 3.3.1  Internal Evaluation

The formative evaluation consists of pre and post exams aimed at testing the basic math skills utilized in the module. Student performance and major changes are monitored in semesters following module participation until graduation. Pre and post surveys are conducted focusing on the improvements needed in implementation.

### 3.3.2  Pre-survey

Pre-surveys were administered in the first week of each module semester (during the introductory meeting) by the module instructor. These surveys gathered information about the demographics of the students registered for the module. In addition, the preliminary surveys gather information on the student's math background and initial perception of uses of mathematics in STEM fields. Instructors are then able to gage the module according to the information obtained.

### 3.3.3  Post-survey

Post-surveys were administered at the last meeting of the module during the semester by the module instructor. These surveys gathered information on the students' satisfaction of the instructors and the material taught. In addition, these surveys gaged the students' perception of how much their participation in the modules helped their performance in their math and biology courses. The information obtained from these surveys was used to make improvements to the modules for future implementation.

### 3.3.4  Follow-Up Surveys

After a student has completed at least a semester of study after the participation in a module, participants will complete an online follow-up survey to gage the retention and usefulness of the knowledge obtained from the modules in the subsequent semesters. Students are asked to participate in these follow-up surveys every semester until graduation.

### 3.3.5   Student Performance and Retention in CES Majors

In the institutions' core calculus courses, semester and final exams are recorded for each student. Of this program comparisons will be made for participants in these modules versus comparable students that did not participate for the semester the student takes the module as well as subsequent semesters. In addition, the participating students will be monitored for change in majors to a major outside of STEM until graduation. Comparisons will be made on proportion of participating students who switch majors (to outside STEM) to a comparable group of students who did not participate in the modules. Comparable students will be obtained via quantitative measures such as math SAT score, previous exam scores, and math placement scores. These student control groups will be chosen with assistance from an in-house statistician, and the identity of these participants will be kept blind from the participating instructors and departments until the conclusion of each semester and module.

### 3.3.6   Exit Interviews

Interviews with all participating students were conducted, with a project member or undergraduate not associated with the module in question, at the end of the semester the module is implemented. The interviews were intended to gather information from students about the implementation of the modules and any concerns or improvement suggestions the students may have. Information obtained from these interviews will be used to improve the modules for future implementation.

## 3.4   Discussion

A key component of this work is the use of multi-departmental (or multi-disciplinary) collaborations to arrive at a greater academic impact. In the case of the authors host institution, collaborative educational activities between the department of Mathematical Sciences and Bioengineering were originally sparked by a creative inquiry project to that focused on bringing undergraduate students and faculty from both departments together to explore research areas and ideas that bridge the disciplines and require the expertise of both fields to address biomedical and applied mathematical concepts. The work presented here is a further extension of this collaboration, and offers a further bridge between the two departments.

Dissemination of this work is a key component of this project. As is the case with the current work, the results of each module will be assembled for conference dissemination. A project web site has been developed that can be used by participating students, and this site will be opened to other institutions to assist in implementing similar programs at their high school, college or university. For each module the web site will house video of lectures, worksheets, podcasts, pictures from field trips, and more. In addition to the module information the web site will have the results of all pre- and post-surveys, follow-up surveys, reports of assessment from each evaluation period, and a final report from the entire 2-year project.

Long-term plans for this work include expansion of the modules to include Data Mining, Genetic Sequencing, Nano-Medicine, BioFluid Dynamics, and Network simulation for the Smart Grid Technologies. Dissemination would hope to expand the program into other institutions. An external evaluation model will be used to assess the implementation of the program, and the final report will be shared and published through the project website and by the project members at various conferences.

The goal of engaging, mentoring and retaining STEM students can be empty rhetoric without a lot of creative thinking. The project presented here stands on the shoulders of creative projects in the Mathematics and Bioengineering departments that involve new instructional methodology, new uses of technology, and experience in creative inquiry connecting undergraduates with experts in various fields in the university and industry. The project also stands on the shoulders of the host institution's efforts at developing undergraduate scholarship in the Creative Inquiry program. The program strategy and plan involves faculty members who have participated in projects like the ones described above and who have experience in the development of applied-learning experiences in Biomedicine that involve quantitative issues at the level of the students' current math courses. The evaluation of the project makes use of the extensive database on individual student performance maintained by the Mathematics Department. The project management team consisted of faculty members from Mathematical Sciences and Bioengineering who were enthusiastic about working together to recruit students for this project, implement the research experiences, accompany students on trips to labs in medicine and industry, mentor the students in their mathematics courses, and evaluate the project.

**Acknowledgements** The authors wish to thank Clemson University for supporting the efforts of the Creative Inquiry program and the NSF DUE Award number 1044265. Any opinions, findings, and conclusions or recommendations expressed in this material are those of the author and do not necessarily reflect the views of the National Science Foundation.

# References

1. Beichner, R.: Student-centered activities for large-enrollment university physics (SCALE-UP), principal investigator, in invention and impact: Building Excellence in Undergraduate Science, Technology, Engineering and Mathematics (STEM) Education, NSF award 9981107, pp. 61–66. AAS (2005)
2. National Science Board: Undergraduate Science, Mathematics and Engineering Education: Role for the National Science Foundation and Recommendations for Action by Other Sectors to Strengthen Collegiate Education and Pursue Excellence in the Next Generation of U.S.

Leadership in Science and Technology, Report of the Task Committee on Undergraduate Science and Engineering Education. Neal, I-I., Chair, Washington, DC (1986)

3. National Science Foundation: Proactive Recruitment in Science and Mathematics, Synopsis of the PRISM Program. Solicitation 09-596 (2008)

4. Ohland, M.W., Sill, B.: Identifying and removing a calculus pre-requisite as a bottleneck in Clemsons general engineering curriculum. J. Eng. Educ. **93**(3), 95–99 (2004)

5. Reba, M., Weaver, B.: Tablet PC-enabled active learning in mathematics: a first study. In: Proceedings of the International Workshop on Pen-Based Learning Technologies, pp. 10–16. IEEE (2007)

6. Schwartz, M., Hazari, Z., Sadler, P.: Divergent voices: views from teachers and professors on pre-college factors that influence college science success. Sci. Educ. **17**(1), 18–35 (2008)

7. Seymour, E., Hewitt, N.: Talking about leaving: why undergraduates leave the sciences. Westview, Boulder (1997)

8. Viktorova, I., Scruggs, M., Zeller, I., Faichild, K.: An analysis of heat explosion for thermally insulated and conducting systems. Appl. Math. **3**(6), 535–540 (2012). ISSN: 2152-7385

# Chapter 4
# Proving the "Proof": Interdisciplinary Undergraduate Research Positively Impacts Students

M.L. Crowe, J. Rychtář, O. Rueppell, M. Chhetri, D.L. Remington, and S.N. Gupta

## 4.1 Introduction

The biological sciences encompass a broad spectrum of academic fields and most sub-disciplines include mathematical modeling and statistical analysis as an integrative component of their scientific process. Advances in computational technology have promoted the growth of the newest interdisciplinary fields such as epidemiology, systems biology, neuroscience, genomics and nanotechnology and bioinformatics. These interdisciplinary areas of study are data rich, requiring new mathematical models and tools to recognize patterns and manage information. The increasingly sophisticated modeling and analytical techniques of these and other biological fields require the twenty-first century biologist to possess more advanced skills in mathematics. Conversely, the most productive contemporary mathematicians have a broad, interdisciplinary scientific training, with most prospects interfacing with the biological sciences. Educational approaches to prepare biology and mathematics students for these twenty-first century career opportunities, however, have lagged behind the recent advances in mathematical and computational applications in biology. The Mathematical Association of America

M.L. Crowe (✉)
Associate Provost of Experiential Education, Florida Southern College,
Lakeland, FL 33801-5698, USA
e-mail: mcrowe@flsouthern.edu

J. Rychtář • M. Chhetri • S.N. Gupta
Department of Mathematics and Statistics, The University of North Carolina at Greensboro,
Greensboro, NC 27402, USA
e-mail: rychtar@uncg.edu; m_chhetr@uncg.edu; sngupta@uncg.edu

O. Rueppell • D.L. Remington
Department of Biology, The University of North Carolina at Greensboro,
Greensboro, NC 27402, USA
e-mail: olav_rueppell@uncg.edu; dlreming@uncg.edu

J. Rychtář et al. (eds.), *Topics from the 8th Annual UNCG Regional Mathematics and Statistics Conference*, Springer Proceedings in Mathematics & Statistics 64, DOI 10.1007/978-1-4614-9332-7_4, © Springer Science+Business Media New York 2013

report Undergraduate Programs and Courses in the Mathematical Sciences: CUPM Curriculum Guide 2004 [2] outlined six approaches to improving campus-wide quantitative literacy, including creating interdisciplinary undergraduate research projects for students. These research projects help students develop quantitative skills that are not often achieved in the traditional classroom setting [7]. Students who are involved in undergraduate research gain self-confidence [5, 11] are more likely to complete their undergraduate education [10, 12] and are more likely to go onto graduate school compared to students who did not have a research experience [1,4,6,12,16,19]. Furthermore, various intellectual gains result from undergraduate research, including critical thinking and problem solving [8, 9, 13, 15, 17, 20]. The benefits of research projects include an increased understanding of content, the ability to explain things to others, in general, improvement of writing and communication skill [3,14,18,21].

## 4.2 MathBio Program at UNCG

In 2006 faculty members within the Department of Biology, Department of Mathematics and Statistics and the Office of Undergraduate Research (OUR) at the University of North Carolina at Greensboro (UNCG) came together to develop a year-long MathBio undergraduate research experience program, sponsored by the National Science Foundation (NSF 0634182; 0926288).

The objectives of the program included:

- generating new knowledge at the interface of mathematics and biology,
- showcasing the importance of mathematics outside the discipline and the use of mathematics and statistics in the field of biology,
- guiding students to an increased proficiency of research skills,
- preparing students for graduate program in biology, mathematics or for any interdisciplinary program.

The primary activity of the program was to involve teams of biology majors and mathematics majors working on interdisciplinary research projects co-mentored by both biology and mathematical science faculty members. Over the course of a 12-month period, each team was to develop a research question, a plan of action, and a timetable to carry out experiments and/or simulations to investigate the questions. Students were involved in every step of the research cycle from synthesizing primary literature, collecting and analyzing data, to presenting the results of their projects. They had to learn new software programs and find new ways of data analysis. Students and faculty mentors participated in the program part-time during the academic year and full-time for 10 weeks in the summer.

## 4.3  Impact of the MathBio program

We have supported the research of eight to nine undergraduate students during each of the past 6 years, with a total of 44 unique students involved in the program. Some of the students participated for multiple years, as we typically attract sophomore or juniors into the program. Forty-eight percent of our participants were been women and 21% were from under-represented groups in STEM (specifically African American). Fifteen faculty members in Biology and Mathematics and Statistics have co-mentored the students. Students gained financial compensation for their involvement in our program while faculty members got a small stipend and funding for laboratory materials, supplies, and software.

### *4.3.1  Impact on Student Participant Post-Baccalaureate Degree Plans*

Thirty-four participants have graduated while ten are still enrolled in their undergraduate degree programs. Twenty of the 34 graduates are enrolled with assistantships in either graduate or MD/PhD programs, while two have already finished their MS degrees (in Computer Sciences and Chemical Engineering) and one already completed a PhD (in Statistics). This is noteworthy because fewer than 30% of UNCG's biology and mathematics majors indicate they plan to continue their education by enrolling in post-graduate study (UNCG Fact Book 2011). Another former participant in our program is now teaching high-school mathematics in a rural county in North Carolina, directly improving STEM education.

### *4.3.2  Impact on Student Participant Learning*

We have analyzed the impact our program has had on student learning by adopting existing public surveys of student self-reported outcomes [17] and administering them to student participants at the end of their involvement in our program. Our participants reported "significant" gains in their ability to ready primary literature, critically analyze information, define and solve problems, and in their ability to communicate in writing. They report "some" gains in their oral communication skills, in their ability to think innovatively, in understanding ethical issues faced by scientists and clarification of a career path. They also report that their writing ability improved as a result of their experience.

### 4.3.3   Impact on Student Participant Professional Development

Our participants have given more than 200 poster and/or oral presentations at regional, national, and international meetings. Thirteen of the participants won awards for outstanding presentations at ten different meetings, including international conferences. The program has resulted in 32 peer-reviewed publications with undergraduates as co-authors in journals such as Journal of Mathematical Biology, Journal of Theoretical Biology, Journal of Evolutionary Ecology, and Journal of Interdisciplinary Mathematics.

### 4.3.4   Outreach

Faculty members and participants interacted with students/teachers by:

1. developing and presenting two full days of mathematical biology curriculum for a summer 4-H camp;
2. developing and presenting hands-on materials for a high-school biology and mathematics courses, a high-school environmental science class, and for an elementary school, and
3. bringing home-schooled elementary students out in the field.

In 2009 another former undergraduate participant who had subsequently become a high-school teacher in the region brought part of his high-school class to the UNCG Mathematics and Statistics conference and the high-school students interacted with our math-bio participants.

## 4.4   Conclusions

The UNCG MathBio program has achieved its objectives of setting undergraduate Biology and Mathematics students on a path toward productive careers as twenty-first century scientists and educators. Moreover, the publications resulting from MathBio projects demonstrate the extent to which undergraduate research can produce genuine scientific advancement. We hope our experience will motivate and encourage others to pursue similar efforts.

**Acknowledgments** This material is based upon work supported by the National Science Foundation under grant numbers DMS 0634182 and DBI 0926288. Any opinions, findings, and conclusions or recommendations expressed in this material are those of the author and do not necessarily reflect the views of the National Science Foundation.

# References

1. Alexander, B.B., Foertsch, J., Daffinrud, S., Tapia, R.: The spend a summer with a scientist (sas) program at rice university: a study of program outcomes and essential elements, 1991–1997. CUR Q. **20**(3), 127–133 (2000)
2. Barker, W., Bressoud, D., Epp, S., Ganter S., Haver B., Pollatsek, H.: Undergraduate Programs and Courses in the Mathematical Sciences: CUPM Curriculum Guide. ERIC (2004)
3. Barratt, N.M.: Field botanist for a day: a group exercise for the introductory botany lab. Am. Biol. Teach. **66**(5), 361–362 (2004)
4. Bauer, K.W., Bennett, J.S.: Alumni perceptions used to assess undergraduate research experience. J. High. Educ. **74**(2), 210–230 (2003)
5. Campbell, A., Skoog, G.: Preparing undergraduate women for science careers: facilitating success in professional research. J. Coll. Sci. Teach. **33**(5), 24–26 (2004)
6. Chandra, U., Stoecklin, S., Harmon, M.: A successful model for introducing research in an undergraduate program. J. Coll. Sci. Teach. **28**, 113–116 (1998)
7. Ganter, S.: Creating networks as vehicles for change. In: Steen, L.A., Madison, B.L. (eds.) The Future of Quantitative Literacy. National Council on Education and the Disciplines, Princeton (2003)
8. Hakim, T.: Soft assessment of undergraduate research: reactions and student perceptions. Counc. Undergrad. Res. Q. **18**, 189–192 (1998)
9. Hathaway, R.S., Nagda, B.R.A., Gregerman, S.R.: The relationship of undergraduate research participation to graduate and professional education pursuit: An empirical study. J. Coll. Stud. Dev. **43**(5), 614–631 (2002)
10. Hippel, W., Lerner, J.S., Gregerman, S.R., Nagda, B.A., Jonides, J.: Undergraduate student-faculty research partnerships affect student retention. Rev. High. Educ. **22**(1), 55–72 (1998)
11. Houlden, R.L., Raja, J.B., Collier, C.P., Clark, A.F., Waugh, J.M.: Medical students' perceptions of an undergraduate research elective. Med. Teach. **26**(7), 659–661 (2004)
12. Ishiyama, J.: Undergraduate research and the success of first-generation, low-income college students. Counc. Undergrad. Res. Q. **22**(1), 36–41 (2001)
13. Ishiyama, J.: Does early participation in undergraduate research benefit social science and humanities students? Coll. Stud. J. **36**(3), 381–387 (2002)
14. Johnson, D.W., Johnson, R.T., Smith, K.A.: Active Learning: Cooperation in the College Classroom. Interactive Book Company, Edina (1998)
15. Kardash, C.A.M.: Evaluation of undergraduate research experience: perceptions of undergraduate interns and their faculty mentors. J. Educ. Psychol. **92**(1), 191 (2000)
16. Kremer, J.F., Bringle, R.G.: The effects of an intensive research experience on the careers of talented undergraduates. J. Res. Dev. Educ. **24**(1), 1–5 (1990)
17. Lopatto, D.: Survey of undergraduate research experiences (sure): first findings. Cell Biol. Educ. **3**(4), 270–277 (2004)
18. Lord, T.R. 101 reasons for using cooperative learning in biology teaching. Am. Biol. Teach. **63**(1), 30–38 (2001)
19. Nnadozie, E., Ishiyama, J.T., Chon, J.: Undergraduate research internships and graduate school success. Journal of College Student Development **42**, 145–156 (2001)
20. Seymour, E., Hunter, A.B., Laursen, S.L., DeAntoni, T.: Establishing the benefits of research experiences for undergraduates in the sciences: first findings from a three-year study. Sci. Educ. **88**(4), 493–534 (2004)
21. Tenney, A., Houck, B.: Learning about leadership: team learning's effect on peer leaders. J. Coll. Sci. Teach. **33**(6), 25–29 (2004)

# Chapter 5
# Modeling Heat Explosion for a Viscoelastic Material

Irina Viktorova, Kyle Fairchild, and Jeff Fischer

## 5.1  Introduction

### 5.1.1  Heat Conduction

Heat conduction is a mechanism of heat transfer occurring through a solid material. The rate equation for heat conduction is known as Fourier's law. Fourier's law defines the heat transfer rate as directly proportional to some spatial temperature difference $\Delta T$. These temperature gradients within the material represent the driving potential for heat propagation. One of the limiting factors of Fourier's law is that it implies infinite speed of heat propagation as well as infinite heat flux for boundary conditions or extremely high rates of temperature change. The Maxwell–Cattaneo equation of heat conduction allows for more apt modeling with respect to problems of large heat fluctuations resulting in hyperbolic equations for heat propagation [1, 2].

### 5.1.2  Heat Explosion

Material failure is a well-researched topic in material science, and although most failure mechanics are observed in terms of crack initiation and subsequent crack propagation, the exact situations determining material failure can become much more complicated. One such complication occurs when the mechanism of loading the material is no longer a static condition but becomes a repeated pattern of loading

I. Viktorova (✉) • K. Fairchild
Department of Mathematical Sciences, Clemson University, Clemson, SC, USA

J. Fischer
Department of Mechanical Engineering, Clemson University, Clemson, SC, USA

J. Rychtář et al. (eds.), *Topics from the 8th Annual UNCG Regional Mathematics and Statistics Conference*, Springer Proceedings in Mathematics & Statistics 64, DOI 10.1007/978-1-4614-9332-7__5, © Springer Science+Business Media New York 2013

and unloading [4]. In the case of polymeric material and composites there are special cases where the viscous resistance of the material can generate an internal thermal energy proportionate to both the magnitude and frequency of loading [5]. Such phenomena have been seen in studies with respect to tension compression testing of glass reinforced plastic [6].

The two primary laws of heat conduction, Fourier's law of heat conduction and Maxwell's heat conduction law, dictate that heat will diffuse proportionally to temperature from high to low concentrations. Under ordinary conditions the thermal energy is dissipated at approximately the same rate at which it is generated. Creating a stationary thermal state, however, in cases where the heat generated is significantly greater than the heat dissipated will lead to a phenomenon known as Heat Explosion.

Heat explosion is a catastrophic failure of the material analogous to what would be expected from the sudden heat flux of an exothermic chemical reaction. The focal point of heat explosion theory is the idea that although mechanical behavior of a material can lead directly to fatigue failure, failure can also occur less intuitively in the form of thermal failure [3].

## 5.1.3 Parameters

Cyclic loading occurs in engineering applications ranging from aviation composite steel to automotive engine walls and artificial knee joints. The ultimate goal in material selection and design for any of these applications is to be able to model and predict the occurrence of thermal failure in the form of heat explosion. In order to do this in increasingly complex systems, it is common practice to simplify the conditions of the system by making assumptions on parameters for both the environment and the material. Although these assumptions make the model more manageable in terms of feasibility and complexity, they inherently detract from the significance and accuracy of the result. For this reason the goal of this paper is to develop a model that can predict heat explosion while limiting assumptions regarding the condition of the system and in doing so, increasing the accuracy and usefulness of the model.

The novel approach regarding the model proposed by this paper lies in its ability to predict thermal failure using material properties, and in doing so limiting the parameters that need to be assumed. This paper elaborates on the connection that can be established between mechanical properties and thermal properties of a material. These properties can be collectively referred to as properties of thermo-viscoelastic parameters. Using standard material creep testing, material specific parameters can be established empirically and applied using the ideas of Fourier's law of heat conduction. Because this model focuses heavily on mechanical properties of a material, it is possible to devise a model that reduces the amount of required assumptions of the system and in doing so the model becomes both more effective and more significant.

There are three main material parameters factored into this model. The material property for heat retained by the system under cyclic loading ($\gamma$), the material property for heat dissipated by the system ($\beta$), and the material property for influence of the heat on the material ($\delta$). There also exists a delta critical ($\delta_*$) which represents a unique condition of $\delta$ at the instant prior to heat explosion [7]. $T$ represents the temperature of the system while $T_m$ is the temperature of the material. Eta ($\eta$) is defined as the ratio of $T$ to $T_m$ and is used in the integration equations [7]. Although these are defined parameters, it is very difficult to give a physical manifestation of their meaning. For the time being these are all represented as dimensionless material parameters that will be given concrete meaning in work to be done in the future [7].

In modeling heat propagation, we will use the Maxwell–Cattaneo relativistic heat equation. The common Laplace operator is given as ($\Delta$), the thermal diffusivity coefficient ($\alpha$), as well as constants for density given as ($\rho$) and specific heat as ($c$). In the heat equation it is important to note that ($T$) represents the temperature gradient, with ($t$) relating to time and the heat release intensity being represented as ($Q$). When considering the relativistic heat equation, the material will have a property called the relaxation time ($\tau$). The relaxation time depends on the ability of the material to recover to an equilibrium position when loads from external sources are removed.

## 5.2 Governing Equations

### 5.2.1 Modeling Equations

The modeling equations for most heat transfer processes can be derived from Fourier's law of Heat Conduction. The equations for this specific study match the Fourier system developed by Viktorova [6].

$$\delta_* = \left\{ \frac{1+\gamma}{2} \left[ T_m^{\frac{1-\gamma}{2}} \int_{\frac{1}{T_m}}^{1} \frac{d\eta}{\sqrt{1-\eta^{1+\gamma}}} \right] \right\}^2 \tag{5.1}$$

This equation is used to find the critical heat influence within a material that causes heat explosion, delta critical. In this case, heat removal is assumed to be zero. This represents a perfectly insulated scenario where no heat is dissipated. The right side of the equation is a Cauchy problem setting in terms of $T_m$. $T_m$ must also satisfy the boundary conditions of the specimen in order to accurately model the physical sample. The Cauchy process is used to find $\delta_*$ as $T_m$ is increased to infinity [7]. The heat influence value rises quickly until the instant of heat explosion and then rapidly declines. For any specimen, heat explosion occurs at only one temperature and that temperature is only dependent on $\gamma$. This is important because it validates comparison when $\beta$ is no longer zero. The next equation models that situation.

$$\delta_* = \left\{ \frac{1+\gamma}{2} \left[ \int_1^{T_m} \frac{dT}{\sqrt{[T_m^{1+\gamma} - T^{1+\gamma}] + [\frac{\beta(1+\gamma)}{2}(T - T_m)(T + T_m - 2)]}} \right]^2 \right\}$$

(5.2)

### 5.2.2  Heat Transfer

The classical defining equation for the rate of heat transfer by conduction is given by Fourier's law. Fourier's law for the one-dimensional plane wall having a temperature distribution $T(x)$ is given by Eq. (5.3), where $k$ is the thermal conductivity of the material, and $q_x''$ is the heat flux, or heat transfer rate per unit area, and $\frac{dT}{dx}$ is the temperature gradient in the $x$ direction [2].

$$q_x'' = -k \frac{dT}{dx}$$

(5.3)

Utilization of Fourier's law is limited by the implication of an infinite rate of heat propagation for extremely high rates of temperature variation. The Maxwell–Cattaneo equation for heat transfer is more suited for modeling heat transfer for condition with high temperature transience such as a series of pulses. This equation is shown below as Eq. (5.4) [6].

$$\tau \frac{\delta^2 T}{\delta t^2} + \frac{\delta T}{\delta t} = \alpha \Delta T + \frac{Q}{\rho c}$$

(5.4)

## 5.3  Results and Discussion

### 5.3.1  Results Based on $\beta$

Figure 5.1 shows $\delta_*$ ratios with respect to $\beta$ over nine different values of $\gamma$. As $\beta$ increases it can be seen that the $\delta_*$ ratio also increases, such that the value of $\delta_*$ with $\beta$ is constantly increasing with respect to $\beta$. In comparing different $\gamma$ values, we can see that at lower levels of $\gamma$, an increase in $\beta$ will have a greater effect on the resulting $\delta_*$ ratio. This is important as materials that relatively retain a lower value of heat will require a much greater heat and thus a greater $\delta_*$ value to undergo heat

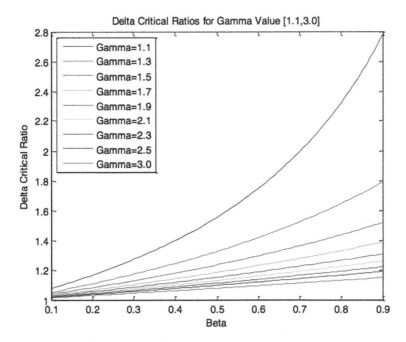

**Fig. 5.1** Delta critical ratios are displayed as beta ranges from 0.1 to 0.9. *Each line* reflects a different gamma value. These values are listed in the legend

explosion as expected. From Fig. 5.1, the ratio of $\delta_*$ with respect to $\beta$ is not linear. This suggests that an increase in heat removal will have a greater increase in delta critical, and thus more heat will be required to experience heat explosion.

### 5.3.2  Effects of $\gamma$ on Delta Critical Ratios

Figure 5.2 depicts how the $\delta_*$ ratios are greater for high values of $\beta$ and low values of $\gamma$. $\delta_*$ ratios are about equally affected by $\gamma$ as they are $\beta$ for our range considered. Given large values of $\gamma$, $\beta$ has little effect on the $\delta_*$ ratio. For small values of $\gamma$, $\beta$ has a great effect on $\delta_*$ ratios. Conversely, $\gamma$ affects $\delta_*$ more greatly for larger values of $\beta$ and less for lower values of $\beta$. Figure 5.2 suggests that for materials that have low heat retention, the effect of heat dissipation greatly affects the heat that is required for heat explosion. For a material that has high heat retention, it is not as important to consider the effects of heat dissipation as the effect on the temperature at which heat explosion occurs.

Considering a situation where the factor of heat removal is considered constant, if the heat removal factor is low, then the temperature at which heat explosions occur does not vary with respect to the heat retention of the material. For high heat removal factors, a small heat retention property in the material used will maximize temperature at which heat explosion will occur.

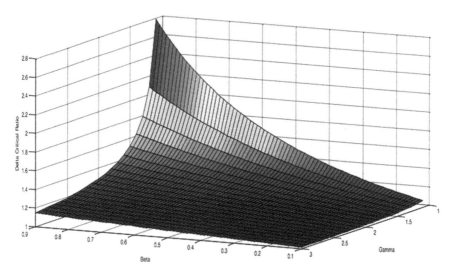

**Fig. 5.2** 3-D rendering of the delta critical ratios displayed in Fig. 5.1

### 5.3.3   Future Work: Mechanical Parameter of Loading

We wish to consider how the cyclic loading of heat will affect a viscoelastic material given the use of the Maxwell–Cattaneo heat equation. The use of the Maxwell–Cattaneo equation allows us to consider boundary conditions caused by high-frequency pulsed heating conditions. When considering the cyclic loading, we wish to vary the frequency as well as the parameters of the amplitude to determine the effects on the overall temperature increase. That is, we wish to determine whether the amplitude or frequency will affect the ability of the heat transfer to increase temperature more rapidly. We wish to perform a sensitivity analysis on the effects of both to determine which has a greater effect in increasing the heat rate. Our current prediction is that the loading frequency will have a greater effect on the overall temperature of the material [6].

## 5.4   Conclusions

This paper presents a mathematical approach to explain the causes of heat explosion. The simplified approach to modeling heat explosion represents a direct comparison between the effects of heat removal and the heat retention of the material. A better understanding of the causes of heat explosion has been achieved, as well as identifying the relative effect of heat removal and heat retention.

The Cauchy problem setting has shown that the material's heat retention rate is about an equivalent factor to the conditions relevant to heat removal. Considering a scenario of an airplane wing where the material property will be held constant due to weight limits, it is important to consider that heat removal will have an influence on the rate of heat explosion. This will allow a company to modify air flow about the wing in order to improve the heat removal rate, and thus increase the heat required for heat explosion to occur. Observing a scenario of a component of an engine block where the boundary conditions are held constant, that is, when the heat removal coefficient is constant, changing the material to be more resistant to heat change will increase the total heat required to enter the system for heat explosion to occur. When a company is experiencing heat explosion in a constant heat removal setting, it is important to consider material changes that would require a greater heat before heat explosion occurs.

Our future goal is to create a reliable model to predict thermal failure on a given geometry for certain material properties with a cyclic loading boundary condition. We wish to model the heat propagation through a material that is undergoing a thermal cyclic load to determine when the material will undergo heat explosion.

**Acknowledgements** We would like to give special thanks to Clemson University and especially the College of Engineering and Science. This project could not have been possible without the contribution of workspace, software licensing and overall support from the faculty.

# References

1. Francis, P.H.: Thermo-mechanical effects in elastic wave propagation: a survey. J. Sound Vib. **21**, 181–192 (1972)
2. Incropera, F.P., Dewitt, D.P., Bergman, T.L., Lavine, A.S.: Fundamentals of Heat and Mass Transfer, 6th edn. Wiley, New York (2007)
3. Meinkohn, D.: Heat explosion theory and vibrational heating of polymers. Int. J. Heat Mass Transfer **25**(4), 645–648 (1981)
4. Oldyrev, P.P.: Heating-up temperature and failure of plastics under cyclic deformation. Mech. Polym. **3**, 483–492 (1967)
5. Oldyrev, P.P., Tamuz, V.P.: Change in properties of glass-reinforced plastic under cyclic tension-compression. Mech. Polym. **5**, 864–872 (1967)
6. Viktorova, I.: The dependence of heat evaluation on parameters of cyclic deformation process. Izv. AN USSR Mech. Solids **4**, 110–114 (1981)
7. Viktorova, I., Suvorova, J.V., Osokin, A.E.: Self-heating of inelastic composites under cyclic deformation. Izv. AN USSR Mech. Solids **19**(1), 516–519 (1984)

# Chapter 6
# Soliton Solutions of a Variation of the Nonlinear Schrödinger Equation

**Erin Middlemas and Jeff Knisley**

## 6.1 Introduction

While linear partial differential equations (PDEs) give rise to low-amplitude waves that occur frequently in the physical world [9], nonlinear waves with nondispersive traits and soliton-like properties can occur naturally also. Soliton-like properties have been observed in water waves, fiber optics, and biological systems such as proteins and DNA [6, 9, 12, 14]. Since linear PDEs fail to take into account phenomena produced by nonlinearity, other mathematical models are needed. Thus, nonlinear PDEs such as the Kortweig de Vries equation and the nonlinear Schrödinger (NLS) equation [5] are used to describe the characteristics of these waves more accurately [2].

Cardiac action potentials (CAPs) also display soliton-like properties. Cardiac cells, like neuron and muscle cells, are excitable cells and are electrically charged by having the membrane act as a capacitor. Previous research [1] has shown CAPs to be well-fit by solutions to the Fitzhugh–Nagumo model,

$$\frac{\partial u}{\partial t} = \frac{\partial^2 u}{\partial x^2} + u(1-u)(a-u) - w \tag{6.1}$$

$$\frac{\partial w}{\partial t} = \varepsilon(u - \gamma w), \tag{6.2}$$

where $u(x,t)$ and $w(x,t)$ are the fast and slow voltage responses at time $t$ and distance $x$ from origin of the CAP, $\gamma$ is the rate of decay of the slow signal when $\varepsilon$ is small so as to model a slower response in $u(x,t)$, and $a$ is the voltage threshold parameter. These two equations account for the discharging of the membrane and

E. Middlemas (✉) • J. Knisley
East Tennessee State University, 807 University Parkway Johnson City, TN 37604, USA
e-mail: zemm16@goldmail.etsu.edu; knisleyj@etsu.edu

J. Rychtář et al. (eds.), *Topics from the 8th Annual UNCG Regional Mathematics and Statistics Conference*, Springer Proceedings in Mathematics & Statistics 64, DOI 10.1007/978-1-4614-9332-7_6, © Springer Science+Business Media New York 2013

the recovery of this charge. If $a = 1$ and $\varepsilon = 0$, the fast voltage response is a traveling wave of the form,

$$u(x,t) = f(x - 2kt) = \frac{1}{1 + Pe^{-(x-2kt)}},$$  (6.3)

where $P$ is a constant term. This fast solution to the Fitzhugh–Nagumo model can also be interpreted as a kink soliton [11]. Due to the characteristics of these traveling waves, there is reason to believe soliton waves that are solutions to a perturbed NLS equation could also describe CAPs.

In this paper, we look into the possibility of CAPs being soliton-like solutions to a perturbed NLS equation. We first determine the perturbed NLS equation that gives rise to solutions describing CAPs. We then study the symmetric properties of the perturbed NLS in order to find more solutions that possibly describe CAPs. To observe if the solutions to our perturbed NLS equation have soliton-like properties, we numerically simulate these solutions.

We discuss our research in the following order. In Sect. 6.2, we introduce background information to solitons and the reasoning behind the methods of our research. In Sect. 6.3, we explain the procedure by which we find our perturbed solution describing CAPs and the perturbed NLS equation. We then introduce the symmetric properties of our perturbed NLS. Section 6.4 provides results and discussion for numerical work. We also conclude with future goals in Sect. 6.4.

## 6.2 Theory

Solitary waves are waves that are localized within a region and retain their form over a certain period of time [14]. These structures have the ability to pass through other waves with only a change of phase. Solitons are solitary waves that are also solutions to completely integrable PDEs. They tend to feature the following properties [14]:

1. They maintain their shape while traveling at a constant speed.
2. They are localized within a region at any given time.
3. They can pass through other waves with no change in amplitude, velocity, or shape.

A particular example of a completely integrable PDE that has soliton solutions is the NLS equation [5]:

$$i\frac{\partial u}{\partial t} = -\frac{\partial^2 u}{\partial x^2} + 2k|u|^2 u.$$  (6.4)

The non-radiating solutions to the NLS are solitons. Due to the interaction of the fast and slow excitation variables within the Fitzhugh–Nagumo model, there is reason to believe that CAPs are soliton-like. If we can show that perturbed solutions of the Fitzhugh–Nagumo model are solutions to a perturbed NLS, then we have evidence to support that CAPs are solitons. Perturbed solutions of the Fitzhugh–Nagumo model can then be used to find a family of closed-form solutions to a Gross–Pitaevskii equation,

$$i\frac{\partial u}{\partial t} = -\frac{\partial^2 u}{\partial x^2} + 2k|u|^2 u + \Phi(x,t,u)u, \tag{6.5}$$

where $\Phi(x,t,u)$ is a potential function. After finding the perturbed solutions for a suitable choice of the potential, a pseudo-spectral method is used to numerically determine the properties of the resulting waves. The closed-form solutions to the Gross–Pitaevskii equation are then utilized to generate more solutions.

## 6.3 Methods

### 6.3.1 Looking for Solitons in a Perturbed NLS

The goal is to find a simple form of $\Phi(x,t,u)$ that allows CAP-like solutions. Motivated by the perturbed Fitzhugh–Nagumo model, we look for solitons in the form of

$$u(x,t) = e^{i\phi}r(x,t),$$

where $\phi = bx + ct$ with $b$ and $c$ as constants [5]. We thus obtain the following:

$$u_t = e^{i\phi}(icr + r_t) \tag{6.6}$$

$$u_x = e^{i\phi}(ibr + r_x) \tag{6.7}$$

$$u_{xx} = e^{i\phi}(-b^2 r + 2ibr_x + r_{xx}). \tag{6.8}$$

We substitute $u_t$, $u_x$, and $u_{xx}$ into Eq. (6.5) and obtain

$$-cr - 2ikr_t = -b^2 r + 2ibr_x + r_{xx} + F(r)r + \Phi(x,t,u)r, \tag{6.9}$$

where $F(r) = 2r^2$. By assumption, $r = f(x - 2kt)$ is a traveling wave with $2k$ being the velocity, from which it follows that

$$r_t = -2kr_x \tag{6.10}$$

Substituting these values of $r_t$ and $r_x$ into Eq. (6.9) suggests the potential in Eq. (6.5) is

$$\Phi(x,t) = k^4 - c - \frac{r_{xx}}{r} - F(r), \tag{6.11}$$

where $c$ is an arbitrary parameter. If $c = k^4$, then Eq. (6.11) implies $\Phi(x, t) = 3|u|u$. Knowing what form of the potential to add to the NLS, we solve for the perturbed solutions of the Fitzhugh–Nagumo model. Since our solution accounts for only the fast variable of the Fitzhugh–Nagumo model, it has infinite energy. To model the fast/slow interaction, we insert a perturbation term $e^{-\delta x}$ for $\delta \approx 0$. This perturbation leads to finite energy solutions. For $a = 1$, our perturbed solutions is a traveling wave of the form,

$$r(x, t) = f(x - 2kt) = \frac{e^{-\delta(x-2kt)}}{1 + Pe^{-(x-2kt)}}. \tag{6.12}$$

Thus,

$$u(x, t) = e^{i\phi} r(x, t) = e^{i\phi} \frac{e^{-\delta(x-2kt)}}{1 + Pe^{-(x-2kt)}} \tag{6.13}$$

is an approximate solution to

$$i \frac{\partial u}{\partial t} = -\frac{\partial^2 u}{\partial x^2} + 2|u|^2 u - 3|u|u. \tag{6.14}$$

### 6.3.2 Symmetries of the NLS Equation

Equation (6.14) is a special case of

$$i \frac{\partial u}{\partial t} = -\frac{\partial^2 u}{\partial x^2} + 2|u|^2 u - M|u|u. \tag{6.15}$$

Indeed, if $M = 3$, we recover (6.14), where if $M = 0$, we obtain the NLS equation. Thus, solutions to (6.14) are not only perturbed solutions to the Fitzhugh–Nagumo model, but may also be closely related to soliton solutions of the NLS. The symmetry group of (6.14), therefore, is a subgroup of the symmetry group of the NLS. The numerical simulations complement several analytic results we have concerning soliton-like properties of CAPs. In particular, for all $M > 3$, Eq. (6.15) admits solutions of the form

$$u(x, t) = e^{i(kx+bt)} r(x - 2kt), \tag{6.16}$$

where $b = -k^2 - 1 + 2M/3$ and where

$$r(x) = \frac{2(2M - 3)e^{1/3 x \sqrt{6M-9}}}{2e^{1/3 x \sqrt{6M-9}} M + M + e^{2/3 x \sqrt{6M-9}} M - 3 - 3e^{2/3 x \sqrt{6M-9}}}. \tag{6.17}$$

These solutions can be extended to larger families of solutions by using Lie symmetry groups, which are subgroups of the permutation groups of the solutions that form smooth manifolds [3, 10]. A Lie group symmetry maps a solution curve to another solution curve. We have shown that Eq. (6.5) is invariant under the following groups:

$$t \to t + t_0, \ x \to x, \ u \to u. \ (time \ translation) \tag{6.18}$$

$$t \to t, \ x \to x + x_0, \ u \to u. \ (spatial \ translation) \tag{6.19}$$

$$t \to t, \ x \to x - ct, \ u \to ue^{i\frac{c}{2}(x-\frac{c}{2}t)}. \ (Galilean \ invariance) \tag{6.20}$$

For example, substituting (6.26) into

$$i\frac{\partial u}{\partial t} = -\frac{\partial^2 u}{\partial x^2} + 2|u|^2 u - M|u|u \tag{6.21}$$

leads to

$$i\frac{\partial(ue^{i\frac{c}{2}(x-\frac{c}{2}t)})}{\partial t} = -\frac{\partial^2(ue^{i\frac{c}{2}(x-\frac{c}{2}t)})}{\partial(x-ct)^2} + 2|ue^{i\frac{c}{2}(x-\frac{c}{2}t)}|^2 ue^{i\frac{c}{2}(x-\frac{c}{2}t)} - M|ue^{i\frac{c}{2}(x-\frac{c}{2}t)}|ue^{i\frac{c}{2}(x-\frac{c}{2}t)}. \tag{6.22}$$

However $|ue^{i\frac{c}{2}(x-\frac{c}{2}t)}| = |u|$. Thus, we can simplify the equation to the following:

$$e^{i\frac{c}{2}(x-\frac{c}{2}t)}i\frac{\partial u}{\partial t} = e^{i\frac{c}{2}(x-\frac{c}{2}t)}\left(-\frac{\partial^2 u}{\partial x^2} + 2|u|^2 u - M|u|u\right). \tag{6.23}$$

The exponential terms cancel. Therefore, the perturbed NLS is Galilean invariant. Determining spatial and temporal symmetries follow the same procedure.

While Eq. (6.14) cannot be solved in closed form except for special cases, we can explore (6.14) numerically to see if its solutions are soliton-like. Specifically, a single soliton retains shape while traveling at a constant speed and also maintains shape when passing through another soliton wave. We will observe these characteristics by looking at not only solutions involving one wave but also solutions involving two waves.

### 6.3.3 The Pseudo-Spectral Method

A pseudo-spectral method is used to numerically solve the Gross–Pitaevskii equation [4, 7, 8]. The method is based on the Fourier transform. If $\int_{-\infty}^{\infty}|f| < \infty$ and $\int_{-\infty}^{\infty}|f|^2 < \infty$, then the Fourier transform exists and is given by

$$\mathfrak{F}(f) = \int_{-\infty}^{\infty} f(x)e^{-2\pi i\omega x}dx. \tag{6.24}$$

It can be shown that

$$\mathfrak{F}\left(\frac{\partial f}{\partial x}\right) = 2\pi i \omega \mathfrak{F}(f). \tag{6.25}$$

Therefore,

$$\frac{\partial f}{\partial x} = \mathfrak{F}^{-1}\left(2\pi i \omega \mathfrak{F}(f)\right). \tag{6.26}$$

Also, it follows that

$$\frac{\partial^2 f}{\partial x^2} = \mathfrak{F}^{-1}\left(-4\pi \omega^2 \mathfrak{F}(f)\right). \tag{6.27}$$

Having our perturbed solutions as initial conditions to the equation, the pseudo-spectral method utilizes special properties of the Fourier transform and its inverses in order to solve the PDE [15]. Beginning with our perturbed NLS,

$$i\frac{\partial u}{\partial t} = -\frac{\partial^2 u}{\partial x^2} + 2|u|^2 u - 3|u|u, \tag{6.28}$$

Equation (6.28) is transformed into

$$i\frac{\partial u}{\partial t} = \mathfrak{F}^{-1}\left(-4\pi \omega^2 \mathfrak{F}(u)\right) + 2|u|^2 u - 3|u|u \tag{6.29}$$

An ode solver is then applied to the resulting ordinary differential equation in order to integrate the solution over a time interval. The solutions are then plotted in order to analyze the soliton-like characteristics of CAPs.

## 6.4 Results

### 6.4.1 Discussion

For Figs. 6.1 and 6.2, CAPs at $\delta = 0.3$ and solutions to our perturbed NLS at $\delta = 0.3$ are compared. Although there is a translational difference within the spatial component between the two waves in Figs. 6.1 and 6.2, there is remarkable similarity of the wave shape between the solution and the actual CAP.

Numerical solutions to (6.15) are computed for different values of $M$ and for the Fitzhugh–Nagumo model initial wave-form. Observing these solutions leads to the following conclusions:

- When $M = 0$, a completely radiating wave is produced.
- When $M = 3$, the wave is non-radiating.
- A smaller value of $M$, however, creates a less dispersive wave.

**Fig. 6.1** Ca cardiac action
potential

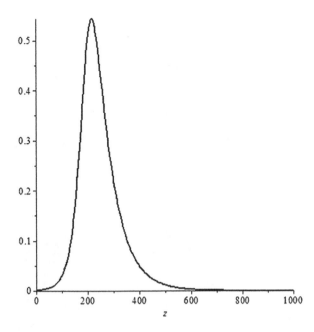

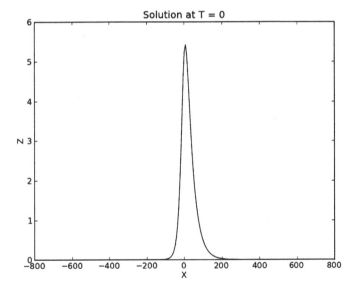

**Fig. 6.2** $\delta = 0.3$, pseudo-spectral method

- For small values of $M$ ($M < 0.3$), the two waves tend to be stationary. This is
  due to the pseudo-spectral method failing to observe the collision between two
  waves at values less than 0.3.

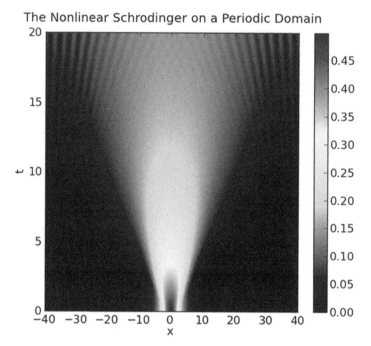

**Fig. 6.3** $M = 0$

For Figs. 6.3 and 6.4, we observe differences between solutions at $\varepsilon = 0$ and solutions at $M = 3$. For $M = 0$, waves immediately radiate as they start to travel, illustrating dispersive properties. At $M = 3$, solutions are hardly radiating. Also, when two solutions of our perturbed NLS collide with each other they maintain their wave-forms and only change by a slight shift in phase, behaving like solitons.

Despite the radiative properties of waves when $M = 0$ and the non-radiative properties of waves when $M = 3$, smaller values of $M$ create less dispersive solutions. This is illustrated with Figs. 6.5 and 6.6. In these two figures, the properties of waves at $M = 1.5$ and waves at $M = 0.3$ are compared. While the wave in Fig. 6.5 disperses significantly after five seconds, the wave in Fig. 6.6 can be considered non-radiating.

We have verified results independently using a central differencing algorithm in the CAS Maple.

### 6.4.2  Future Work

Analyzing the soliton-like properties of CAPs numerically is still in process. Different parameters, such as $M$ values within the Gross–Pitaevskii equation, wave velocities, and $\delta$ terms within the solution from the Fitzhugh–Nagumo model still

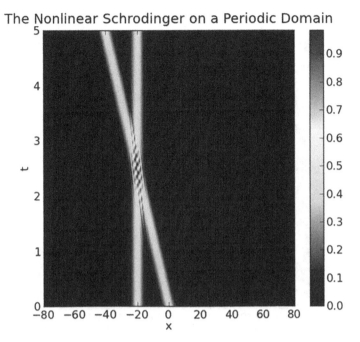

**Fig. 6.4** $M = 3$

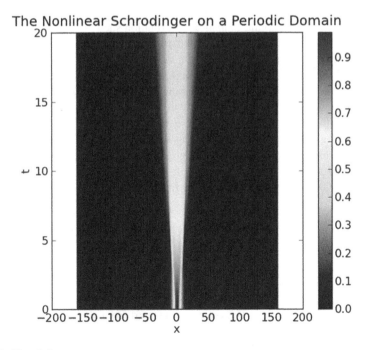

**Fig. 6.5** $M = 1.5$

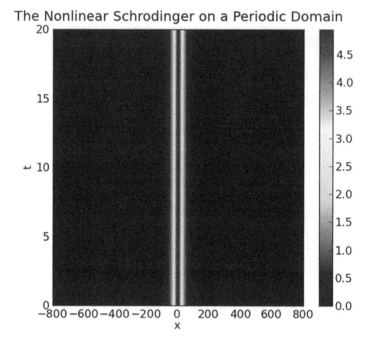

**Fig. 6.6** $M = 0.3$

need to be taken into account. Statistical information such as confidence intervals, standard deviations, and variances of the results from numerical methods will soon be calculated. However, due to errors stated previously, a new computational method, such as the imaginary-time evolution method, will be adopted in the near future. Also, the family of solutions describing CAPs will be discovered by working with more Lie symmetry groups.

There is also value in extending these results to higher spatial dimensions. Although not all the methods for working with one-dimensional NLS equations translate into higher dimensional NLS contexts, some of the results in this paper should extend to higher dimensional settings. Thus, we will also explore higher dimensional, CAP-like solitons in the near future.

## Appendix

The following is our code utilizing the pseudo-spectral method in order to solve our perturbed NLS equation. The code was adapted from a Scipy Cookbook KdV example [13].

```python
import numpy as np
from scipy.integrate import odeint
from scipy.fftpack import diff as psdiff

#from mpl_toolkits.mplot3d import Axes3D
from matplotlib.collections import PolyCollection
from matplotlib.colors import colorConverter

#from mpl_toolkits.mplot3d import axes3d
import matplotlib.pyplot as plt


def shr_exact(x, c):
    """Profile of the exact solution to the KdV for a
    single soliton on the real line."""
    #u = 1.2*1/(np.cosh(1.2*(x+20)))
    #    +np.exp(8j*(x))*0.8*1/(np.cosh(.8*x))
    eps = 2.0

    delta = 0.8

    beta = eps

    gamma = 1/eps

    u = (np.exp(-delta*x))/(1+np.exp(-x))
        +(np.exp(-delta*(x+20)))/(1+np.exp(-(x+20)))
        *np.exp(3j*(x))
    # u = np.exp(-delta*(beta*(x)))*np.exp(0j*(x))/
    #     (1+np.exp(-(beta*(x))))
    #u = np.exp(-delta*(beta*x))/(1+np.exp(-(beta*x)))
    u = gamma*u
    u = np.array(u, dtype=np.complex64)
    u = np.array([u.real, u.imag])
    u = u.flatten()

    return u

def shr(u, t, L):
    """Differential equations for the KdV equation,
    discretized in x."""
    # Compute the x derivatives using the
    pseudo-spectral method.
    # ux = psdiff(u, period=L)
```

```
eps = 2.0
gamma = 1/eps
n = len(u)
uxxRe = psdiff(u[0:(n/2)], period=L, order=2)
uxxIm = psdiff(u[(n/2):n], period=L, order=2)

uxx = np.array([uxxRe,uxxIm])
uxx = uxx.flatten()

absu =np.sqrt(u[0:n/2]**2+u[n/2:n]**2)
absu = np.array([absu,absu])
absu = absu.flatten()

absu2 = u[0:n/2]**2+u[n/2:n]**2
absu2 = np.array([absu2,absu2])
absu2 = absu2.flatten()

# Compute du/dt = -i*( -uxx - 2abs(u)u )
= i * (uxx + 2abs(u)u)
dudt = (-1*2*absu2)*u + uxx + eps*(3*absu)*u
idudt= np.array([-1*dudt[(n/2):n], dudt[0:(n/2)]])
return idudt.flatten()
#return ( idudt.real, idudt.imag )
```

```
# Set the size of the domain, and create the
discretized grid.
eps = 2.0
beta = eps
L =160.0/beta
N = 256
dx = L/N
x = np.linspace(-L/2, L/2, N)
x1 = np.linspace(-L/beta, L/beta, N)
```

```
# Set the initial conditions.
# Not exact for two solitons on a periodic domain, but
close enough...
u0 = shr_exact(x, 0.75) # + kdv_exact(x-0.65*L, 0.4)
```

```
# Set the time sample grid.
#ps = .01
#alpha = eps**2
Tm = 7
t = np.linspace(0, Tm, 1000)
```

```python
#t = alpha*t

print "Computing the solution."
from mpl_toolkits.mplot3d import Axes3D
from matplotlib.collections import PolyCollection
from matplotlib.colors import colorConverter
sol = odeint(shr, u0, t, args=(L,), mxstep=500)
sol = sol[:, 0:N] + 1j*sol[:,N:(2*N) ]

print "IMshow."

plt.figure(figsize=(6,5))
plt.imshow(np.abs(sol[::-1, :]), extent=[-L/2,L/2,0,Tm])
plt.colorbar()
plt.xlabel('x')
plt.ylabel('t')
plt.axis('normal')
plt.title('The Nonlinear Schrodinger on a Periodic
Domain')
#plt.show()

#print "Wireframe."

#fig = plt.figure()
#ax = fig.add_subplot(111, projection='3d')
#tind = range(0,len(t),10)
#xind = range(0,len(x),5)
#tt = t[tind]
#xx = x[xind]
#ux = abs(sol)[:,xind]
#uu = ux[tind,:]
#X,T = np.meshgrid(xx, tt)
#ax.plot_wireframe(X,T,uu )

#plt.show()

print("WaterFall.")

## Redo the sampling
tind = range(0,len(t),30)
xind = range(0,len(x),1)
tt = t[tind]
xx = x[xind]
ux = abs(sol)[:,xind]
```

```python
#The figure
fig = plt.figure()
ax = fig.gca(projection='3d')

cc = lambda arg: colorConverter.to_rgba(arg, alpha=0.6)

verts = []
for i in tind:
    verts.append( zip(xx,ux[i,:]) )


poly = PolyCollection(verts, facecolors = [cc('b')])
poly.set_alpha(0.3)
ax.add_collection3d(poly, zs=tt, zdir='y')

ax.set_xlabel('X')
ax.set_xlim3d(-L/2,L/2)
ax.set_ylabel('t')
ax.set_ylim3d(0,Tm)
ax.set_zlabel('Z')
ax.set_zlim3d(0,1.1*np.max(abs(sol)))
plt.title('The Nonlinear Schrodinger on a Periodic
Domain')
plt.show()

plt.figure()
plt.plot(xx, abs(sol[0]))
plt.xlabel('X')
plt.ylabel('Z')
plt.title('Solution at T = 0')
plt.show()

plt.figure()
plt.plot(xx, abs(sol[999]))
plt.xlabel('X')
plt.ylabel('Z')
plt.title('Solution at Max Time')
plt.show()

Diff = np.max(abs(sol[0]))-np.max(abs(sol[999]))
Diff = abs(Diff)
print Diff
```

# References

1. Brooks, J.: A singular perturbation approach to the Fitzhugh-Nagumo PDE for modeling cardiac action potentials. Masters thesis, East Tennessee State University. E-thesis—http://libraries.etsu.edu/record=b2340298 S1a (2011)
2. Drazin, P., Johnson, R.: Solitons: An Introduction, p. 15. Cambridge University Press, New York (1989)
3. Gagnon, L., Winternitz, P.: Lie symmetries of a generalised non-linear Schrodinger equation: I. The symmetry group and its subgroups. J. Phys. A Gen. **21**, 1493–1511 (1988)
4. Gottlieb, D., Orzag, S.: Numerical Analysis of Spectral Methods: Theory and Applications. Society for Industrial and Applied Mathematics, Philadelphia (2011)
5. Grimshaw, R., Khusnutdinova, K.: Nonlinear waves, Lecture Notes for MAGIC:Nonlinear Waves (MAGIC021), Birmingham, England (2011)
6. Hasegawa, A., Tappert, F.: Transmission of stationary nonlinear optical pulses in dispersive dielectric fibers. I. Anomalous dispersion. Appl. Phys. Lett. **23**(3), 142–144 (1973) doi:10.1063/1.1654836
7. Hornikx, M., Waxler, R.: The extended Fourier pseudospectral time-domain method for atmospheric sound propagation. J. Acoust. Soc. Am. **128**(4), 1632–1646 (2010)
8. Huang, X., Zhang, X.: A Fourier pseudospectral method for some computational aeroacoustics problems. Aeroacoustics **5**(3), 279–294 (2006)
9. Infeld, E., Rowlands, G.: Nonlinear Waves, Solitons, and Chaos. Cambridge University Press, New York (1990)
10. Popovychh, R.O., Eshraghi, H.: Admissible Point Transformations of Nonlinear Schrodinger 366 Equations, Proceedings of 10th International Conference in MOdern GRoup ANalysis, Larnaca, 167–174 (2005)
11. Remoissenet, M.: Waves Called Solitons, Concepts and Experiments. Springer, Berlin (1999)
12. Sinkala, Z.: Soliton/exciton transport in proteins. Theor. Biol. **241**(4), 919–927 (2006) doi:10.1016/j.jtbi.2006.01.028
13. Weckesseer, W.: Cookbook/KdV. Retrieved on November 24, 2012, from http://www.scipy.org/Cookbook/KdV (22 February 2003)
14. Yakushevich, L.V.: Nonlinear Physics of DNA, 2nd rev. edn. Wiley, Garching (2004)
15. Yang, J.: Nonlinear Waves in Integrable and Nonintegrable Systems. Society for Industrial and Applied Mathematics, Philadelphia (2010)

# Chapter 7
# Galois Groups of 2-Adic Fields of Degree 12 with Automorphism Group of Order 6 and 12

Chad Awtrey and Christopher R. Shill

## 7.1 Introduction

The $p$-adic numbers $\mathbf{Q}_p$ are foundational to much of the twentieth and twenty-first century number theory (e.g., number fields, elliptic curves, $L$-functions, and Galois representations) and are connected to many practical applications in physics and cryptography. Of particular interest to number theorists is the role they play in computational attacks on certain unsolved questions in number theory, such as the Riemann Hypothesis and the Birch and Swinnerton-Dyer conjecture (among others). The task of classifying $p$-adic fields therefore has merit, since the outcomes of such a pursuit can provide computational support to the aforementioned problems as well as other number-theoretic investigations.

Classifying extensions of $\mathbf{Q}_p$ entails gathering explicit data that uniquely determine the extensions, including

1. The number of nonisomorphic extensions for a given prime $p$ and degree $n$ (necessarily finite [15, p. 54]),
2. Defining polynomials for each extension, and
3. The Galois group of the extension's polynomial (a difficult computational problem in general).

When $p \nmid n$ (i.e., tamely ramified extensions) or when $p = n$, then items (1)–(3) are well understood (cf. [1, 12]). When $p \mid n$ and $n$ is composite, the situation is more complicated.

C. Awtrey (✉)
Elon University, Campus Box 2320, Elon, NC 27244, USA
e-mail: cawtrey@elon.edu

C.R. Shill
Elon University, Campus Box 9017, Elon, NC 27244, USA
e-mail: cshill@elon.edu

J. Rychtář et al. (eds.), *Topics from the 8th Annual UNCG Regional Mathematics and Statistics Conference*, Springer Proceedings in Mathematics & Statistics 64, DOI 10.1007/978-1-4614-9332-7_7, © Springer Science+Business Media New York 2013

In this paper, we study items (1)–(3) for degree 12 extensions of $\mathbf{Q}_2$, as extensions of smaller degree have already been discussed in the literature [3–6, 11–13]. Specifically, we focus on Galois extensions as well as those extensions whose automorphism groups have order 6. After describing the computation of defining polynomials of such extensions in the next section, we use the final sections of the paper to show that the Galois groups of these polynomials can be computed solely by knowing the Galois groups of their proper subfields. This approach is of interest, since it offers a method for computing Galois groups of local fields that is different from both the resolvent approach [10, 23, 24] and the Newton polygon approach [9, 19].

## 7.2  The Number of Extensions and Defining Polynomials

In regard to counting the number of extensions of $p$-adic fields, some authors have developed what are known as "mass" formulas [14, 18, 21], where the mass of an extension $K/\mathbf{Q}_p$ takes into account the degree of the extension as well as its automorphism group. The mass is defined as:

$$\text{mass}(K/\mathbf{Q}_p) = \frac{[K : \mathbf{Q}_p]}{|\text{Aut}(K/\mathbf{Q}_p)|}.$$

The mass formulas previously mentioned compute the total mass for all extensions of $\mathbf{Q}_p$ of a given degree. As such, different embeddings are counted separately. Therefore these formulas do not give the number of nonisomorphic extensions. Since there is currently no known formula for computing the number of nonisomorphic extensions of $\mathbf{Q}_p$ for a given degree, the approach taken in the literature is to resolve item (1) by first completing item (2) (cf. [4, 11–13]).

The most general reference for the computation of defining polynomials of $p$-adic fields is [18]. Using the methods of Krasner [14], Pauli–Roblot develop an algorithm for computing extensions of a $p$-adic field of a given degree by providing a generating set of polynomials to cover all possible extensions. Essential to their method is Panayi's root-finding algorithm [16], which can be used to determine whether two polynomials define isomorphic $p$-adic fields.

Table 7.1 shows the number of nonisomorphic extensions of $\mathbf{Q}_p$ of degree $n$ where $p \mid n$ and $n \leq 12$ is composite. This data can be verified by [17], which includes an implementation of the Pauli–Roblot algorithm in its latest release.

**Table 7.1** Number of certain nonisomorphic degree $n$ extensions of $\mathbf{Q}_p$

| $(p,n)$ | $(2,4)$ | $(2,6)$ | $(3,6)$ | $(2,8)$ | $(3,9)$ | $(2,10)$ | $(5,10)$ | $(2,12)$ | $(3,12)$ |
|---------|---------|---------|---------|---------|---------|----------|----------|----------|----------|
| #       | 59      | 47      | 75      | 1,834   | 795     | 158      | 258      | 5,493    | 785      |

**Table 7.2** Polynomials for all degree 12 Galois extensions of $\mathbf{Q}_2$, including ramification index $e$, residue degree $f$, and discriminant exponent $c$

| | Polynomial | $e$ | $f$ | $c$ |
|---|---|---|---|---|
| 1 | $x^{12} + x^6 + x^4 - x + 1$ | 1 | 12 | 0 |
| 2 | $x^{12} - x^{10} - 6x^8 - x^6 + 2x^4 + 7x^2 + 5$ | 3 | 4 | 8 |
| 3 | $x^{12} - 78x^{10} - 1621x^8 + 460x^6 - 1977x^4 + 866x^2 + 749$ | 2 | 6 | 12 |
| 4 | $x^{12} - 162x^{10} + 26423x^8 + 125508x^6 - 64481x^4 - 122498x^2 - 86071$ | 2 | 6 | 12 |
| 5 | $x^{12} - 16x^{10} + 24x^6 + 64x^4 + 64$ | 2 | 6 | 18 |
| 6 | $x^{12} + 52x^{10} - 28x^8 + 8x^6 + 64x^4 - 32x^2 + 64$ | 2 | 6 | 18 |
| 7 | $x^{12} - 156x^{10} + 9900x^8 - 61856x^6 + 33904x^4 + 27712x^2 + 47936$ | 2 | 6 | 18 |
| 8 | $x^{12} - 52x^{10} + 1100x^8 - 12000x^6 - 61072x^4 + 62144x^2 - 62144$ | 2 | 6 | 18 |
| 9 | $x^{12} + 12x^{10} + 12x^8 + 8x^6 + 32x^4 - 16x^2 + 16$ | 6 | 2 | 16 |
| 10 | $x^{12} + x^{10} + 6x^8 - 3x^6 + 6x^4 + x^2 - 3$ | 6 | 2 | 16 |
| 11 | $x^{12} - 84x^{10} + 444x^8 + 32x^6 - 272x^4 - 320x^2 + 64$ | 6 | 2 | 22 |
| 12 | $x^{12} - 60x^6 + 52$ | 6 | 2 | 22 |
| 13 | $x^{12} + 2x^{10} + 4x^8 + 4x^6 + 4x^4 + 4$ | 6 | 2 | 22 |
| 14 | $x^{12} - 20x^6 + 20$ | 6 | 2 | 22 |
| 15 | $x^{12} - 4x^{11} - 10x^{10} + 16x^9 - 6x^8 + 16x^7 + 4x^6 - 8x^5 + 16x^4$ $+ 16x^3 + 16x^2 + 8$ | 4 | 3 | 24 |
| 16 | $x^{12} + 28x^{11} - 2x^{10} + 16x^9 + 26x^8 + 8x^7 + 20x^6 - 24x^5 - 8x^4$ $+ 32x^3 + 32x^2 + 32x + 24$ | 4 | 3 | 24 |
| 17 | $x^{12} + 32x^{11} - 10x^{10} + 8x^9 - 18x^8 + 32x^7 + 20x^6 + 24x^5 - 24x^4$ $+ 32x^3 + 16x^2 - 24$ | 4 | 3 | 24 |
| 18 | $x^{12} - 4x^{11} + 14x^{10} + 36x^9 - 34x^8 - 32x^7 - 48x^6 - 32x^5 + 36x^4$ $- 16x^3 - 40x^2 - 48x + 56$ | 4 | 3 | 24 |
| 19 | $x^{12} - 2x^{11} + 6x^{10} + 4x^9 + 6x^8 + 12x^7 - 4x^6 - 8x^3 + 16x^2 - 8$ | 4 | 3 | 18 |
| 20 | $x^{12} - 8x^{10} - 28x^8 + 40x^6 - 44x^4 + 48x^2 + 40$ | 4 | 3 | 33 |
| 21 | $x^{12} + 8x^{10} - 12x^8 - 24x^6 + 20x^4 - 16x^2 - 24$ | 4 | 3 | 33 |
| 22 | $x^{12} - 8x^{10} - 28x^8 - 8x^6 + 20x^4 + 16x^2 - 24$ | 4 | 3 | 33 |
| 23 | $x^{12} + 4x^{10} + 10x^8 - 8x^6 + 8x^4 + 32x^2 + 8$ | 4 | 3 | 33 |
| 24 | $x^{12} - 24x^{10} + 52x^8 - 8x^6 + 20x^4 + 16x^2 + 40$ | 4 | 3 | 33 |
| 25 | $x^{12} + 28x^{10} - 6x^8 + 40x^6 - 56x^4 - 32x^2 - 56$ | 4 | 3 | 33 |
| 26 | $x^{12} - 4x^{10} + 26x^8 + 8x^6 - 24x^4 + 32x^2 + 8$ | 4 | 3 | 33 |
| 27 | $x^{12} + 36x^{10} + 42x^8 - 40x^6 + 40x^4 + 32x^2 - 56$ | 4 | 3 | 33 |

Using the Pauli–Roblot algorithm [18], we see there are 5,493 degree 12 extensions of $\mathbf{Q}_2$. Using Panayi's root-finding algorithm to compute the size of each extension's automorphism group, we can show that 27 are Galois extensions and 55 have an automorphism group of order 6. For convenience, Tables 7.2 and 7.3 give sample defining polynomials for these two cases, respectively, along with the ramification index, residue degree, and discriminant exponent of the corresponding extension field.

**Table 7.3** Polynomials for all degree 12 extensions of $\mathbf{Q}_2$ that have an automorphism group of order 6, including ramification index $e$, residue degree $f$, and discriminant exponent $c$

| | Polynomial | $e$ | $f$ | $c$ |
|---|---|---|---|---|
| 1 | $x^{12} - 52x^{10} + 20x^8 - 60x^6 - 32x^4 - 16x^2 - 48$ | 3 | 4 | 8 |
| 2 | $x^{12} + 80x^{10} + 81x^8 - 160x^6 - 117x^4 + 80x^2 + 227$ | 2 | 6 | 12 |
| 3 | $x^{12} - 100x^{10} - 59x^8 + 104x^6 + 387x^4 + 444x^2 + 439$ | 2 | 6 | 12 |
| 4 | $x^{12} - 200x^{10} + 7956x^8 - 7360x^6 + 6192x^4 - 2176x^2 - 4672$ | 2 | 6 | 18 |
| 5 | $x^{12} - 864x^{10} - 9916x^8 + 11008x^6 + 14512x^4 + 2560x^2 + 14528$ | 2 | 6 | 18 |
| 6 | $x^{12} - 108x^{10} - 171x^8 + 344x^6 - 61x^4 + 468x^2 + 359$ | 6 | 2 | 16 |
| 7 | $x^{12} - 30x^{10} - 5x^8 + 19x^4 + 30x^2 + 1$ | 6 | 2 | 16 |
| 8 | $x^{12} - 3x^{10} + 4x^8 - 3x^6 + 4x^4 + x^2 + 3$ | 6 | 2 | 16 |
| 9 | $x^{12} + 5x^{10} + 4x^8 + x^6 + 4x^4 + x^2 + 3$ | 6 | 2 | 16 |
| 10 | $x^{12} - 12x^{10} + x^8 + 12x^6 + 15x^4 + 16x^2 + 15$ | 6 | 2 | 16 |
| 11 | $x^{12} + 7x^{10} + 4x^8 + 3x^6 - 4x^4 - x^2 - 5$ | 6 | 2 | 16 |
| 12 | $x^{12} + 20x^{10} - 44x^8 - 4x^6 - 16x^4 - 48$ | 6 | 2 | 16 |
| 13 | $x^{12} + 4x^{10} + x^8 + 4x^6 - x^4 + 8x^2 - 1$ | 6 | 2 | 16 |
| 14 | $x^{12} + 10x^6 + 12$ | 6 | 2 | 22 |
| 15 | $x^{12} + 2x^6 + 4$ | 6 | 2 | 22 |
| 16 | $x^{12} - 2x^6 + 4$ | 6 | 2 | 22 |
| 17 | $x^{12} + 14x^6 - 12$ | 6 | 2 | 22 |
| 18 | $x^{12} - 14x^6 - 12$ | 6 | 2 | 22 |
| 19 | $x^{12} + 12$ | 6 | 2 | 22 |
| 20 | $x^{12} + 14x^6 + 12$ | 6 | 2 | 22 |
| 21 | $x^{12} + 8x^6 - 4$ | 6 | 2 | 22 |
| 22 | $x^{12} - 6x^6 - 4$ | 6 | 2 | 22 |
| 23 | $x^{12} - 2x^6 - 4$ | 6 | 2 | 22 |
| 24 | $x^{12} - 4x^{11} + 10x^{10} - 6x^8 - 8x^7 + 12x^6 + 8x^5 + 8x^4 + 16x^2 + 8$ | 4 | 3 | 24 |
| 25 | $x^{12} + 12x^{11} - 4x^{10} + 4x^9 - 12x^8 + 4x^6 - 8x^5 - 4x^4 + 16x^3 + 8x^2 + 16x - 8$ | 4 | 3 | 24 |
| 26 | $x^{12} + 12x^{10} - 8x^8 + 12x^6 + 4x^4 - 8x^2 + 8$ | 4 | 3 | 27 |
| 27 | $x^{12} + 16x^{10} + 8x^8 + 4x^6 - 12x^4 - 24x^2 - 24$ | 4 | 3 | 27 |

| | | | | |
|---|---|---|---|---|
| 28 | $x^{12} + 32x^{10} + 32x^8 - 4x^6 + 20x^4 + 8x^2 + 24$ | 4 | 3 | 27 |
| 29 | $x^{12} + 16x^{10} + 16x^8 - 4x^6 - 12x^4 + 8x^2 - 8$ | 4 | 3 | 27 |
| 30 | $x^{12} + 8x^{10} + 16x^8 - 4x^6 - 12x^4 + 8x^2 - 8$ | 4 | 3 | 27 |
| 31 | $x^{12} - 20x^{10} + 32x^8 - 12x^6 - 28x^4 - 8x^2 + 24$ | 4 | 3 | 27 |
| 32 | $x^{12} + 4x^{10} - 8x^8 + 12x^6 + 4x^4 - 8x^2 + 8$ | 4 | 3 | 27 |
| 33 | $x^{12} + 20x^{10} - 24x^8 - 4x^6 + 4x^4 - 8x^2 - 24$ | 4 | 3 | 27 |
| 34 | $x^{12} + 6x^{11} + 8x^{10} - 52x^9 - 10x^8 + 24x^7 + 8x^6 + 64x^5 + 28x^4 - 40x^3 - 16x^2 - 16x + 40$ | 4 | 3 | 18 |
| 35 | $x^{12} + 12x^{11} + 8x^{10} + 4x^9 + 16x^8 - 12x^7 - 8x^6 + 8x^5 - 12x^4 + 16x^3 - 8$ | 4 | 3 | 18 |
| 36 | $x^{12} + 2x^{10} - x^8 + 2x^6 + 6x^4 - 4x^2 - 5$ | 4 | 3 | 30 |
| 37 | $x^{12} + 2x^{10} - 11x^8 + 20x^6 + 31x^4 - 30x^2 - 5$ | 4 | 3 | 30 |
| 38 | $x^{12} + 2x^{10} - x^8 - 2x^6 + 2x^4 - 1$ | 4 | 3 | 30 |
| 39 | $x^{12} + 10x^{10} - 99x^8 + 68x^6 + 79x^4 + 74x^2 + 67$ | 4 | 3 | 30 |
| 40 | $x^{12} - 62x^{10} + 33x^8 + 948x^6 + 775x^4 + 162x^2 + 951$ | 4 | 3 | 30 |
| 41 | $x^{12} + 1858x^{10} + 1509x^8 - 1436x^6 + 2047x^4 + 786x^2 + 203$ | 4 | 3 | 30 |
| 42 | $x^{12} + 18x^{10} + 17x^8 - 28x^6 - 57x^4 + 34x^2 + 39$ | 4 | 3 | 30 |
| 43 | $x^{12} - 38x^{10} - 87x^8 + 20x^6 - 41x^4 + 74x^2 + 95$ | 4 | 3 | 30 |
| 44 | $x^{12} + 24x^{10} - 4x^8 - 28x^4 + 32x^2 + 24$ | 4 | 3 | 33 |
| 45 | $x^{12} + 18x^8 - 56x^4 + 40$ | 4 | 3 | 33 |
| 46 | $x^{12} + 8x^{10} + 28x^8 + 24x^6 + 20x^4 - 16x^2 + 24$ | 4 | 3 | 33 |
| 47 | $x^{12} - 12x^{10} + 6x^8 - 24x^6 - 24x^4 + 32x^2 - 8$ | 4 | 3 | 33 |
| 48 | $x^{12} - 8x^{10} - 28x^8 + 4x^4 + 32x^2 + 8$ | 4 | 3 | 33 |
| 49 | $x^{12} + 24x^{10} - 12x^8 + 64x^6 + 4x^4 + 32x^2 - 56$ | 4 | 3 | 33 |
| 50 | $x^{12} - 24x^{10} - 10x^8 - 16x^6 + 8x^4 - 64x^2 + 56$ | 4 | 3 | 33 |
| 51 | $x^{12} - 14x^8 - 24x^4 - 24$ | 4 | 3 | 33 |
| 52 | $x^{12} + 6x^8 + 8x^4 - 8$ | 4 | 3 | 33 |
| 53 | $x^{12} + 8x^{10} - 4x^8 + 48x^6 - 28x^4 - 40$ | 4 | 3 | 33 |
| 54 | $x^{12} + 28x^{10} + 22x^8 + 24x^6 - 24x^4 + 32x^2 - 8$ | 4 | 3 | 33 |
| 55 | $x^{12} + 8x^{10} + 28x^8 - 24x^6 + 20x^4 + 16x^2 + 24$ | 4 | 3 | 33 |

## 7.3  Possible Galois Groups

Having computed a defining polynomial for each extension under consideration, we now turn our attention to determining the Galois group of each polynomial.

Given one of the polynomials $f$ in either Table 7.2 or 7.3, let $K$ denote the corresponding extension defined by adjoining to $\mathbf{Q}_p$ a root of $f$. We wish to compute the Galois group $G$ of $f$, or equivalently the Galois group of the normal closure of $K$. Since the elements of $G$ act as permutations on the roots of $f$, once we fix an ordering on the roots, $G$ can be considered as a subgroup of $S_{12}$, well defined up to conjugation (different orderings correspond to conjugates of $G$). Since the polynomial $f$ is irreducible, $G$ is a transitive subgroup of $S_{12}$; i.e., there is a single orbit for the action of $G$ on the roots of $f$ (each orbit corresponds to an irreducible factor of $f$). Therefore $G$ must be a transitive subgroup of $S_{12}$. Our method for computing Galois groups thus relies on the classification of the 301 transitive subgroups of $S_{12}$ [20].

However, not all of these 301 groups can occur as the Galois group of a degree 12 2-adic field, as we show next.

**Definition 1.** Let $L/\mathbf{Q}_p$ be a Galois extension with Galois group $G$. Let $v$ be the discrete valuation on $L$ and let $\mathbf{Z}_L$ denote the corresponding discrete valuation ring. For an integer $i \geq -1$, we define the **i-th ramification group** of $G$ to be the following set

$$G_i = \{\sigma \in G : v(\sigma(x) - x) \geq i + 1 \text{ for all } x \in \mathbf{Z}_L\}.$$

The ramification groups define a sequence of decreasing normal subgroups which are eventually trivial and which give structural information about the Galois group of a $p$-adic field. A proof of the following result can be found in [22, Chap. 4].

**Lemma 1.** *Let $L/\mathbf{Q}_p$ be a Galois extension with Galois group $G$, and let $G_i$ denote the $i$-th ramification group. Let $\mathfrak{p}$ denote the unique maximal ideal of $\mathbf{Z}_L$ and $U_0$ the units in $L$. For $i \geq 1$, let $U_i = 1 + \mathfrak{p}^i$.*

(a) *For $i \geq 0$, $G_i/G_{i+1}$ is isomorphic to a subgroup of $U_i/U_{i+1}$.*
(b) *The group $G_0/G_1$ is cyclic and isomorphic to a subgroup of the group of roots of unity in the residue field of $L$. Its order is prime to $p$.*
(c) *The quotients $G_i/G_{i+1}$ for $i \geq 1$ are abelian groups and are direct products of cyclic groups of order $p$. The group $G_1$ is a $p$-group.*
(d) *The group $G_0$ is the semi-direct product of a cyclic group of order prime to $p$ with a normal subgroup whose order is a power of $p$.*
(e) *The groups $G_0$ and $G$ are both solvable.*

Applying this lemma to our scenario, where the polynomial $f$ is chosen from Table 7.2 or 7.3, $K/\mathbf{Q}_2$ is the extension defined by $f$, and $G$ is the Galois group of $f$, we see that $G$ is a solvable transitive subgroup of $S_{12}$; of which there are 265 [20]. Furthermore, $G$ contains a solvable normal subgroup $G_0$ such that $G/G_0$ is cyclic of order dividing 12. The group $G_0$ contains a normal subgroup $G_1$ such that $G_1$ is a 2-group (possibly trivial), and $G_0/G_1$ is cyclic of order dividing $2^{[G:G_0]} - 1$. Only 134 subgroups have the correct filtration. Moreover, since the automorphism group of $K/\mathbf{Q}_2$ is isomorphic to the centralizer of $G$ in $S_{12}$, we need to only consider those subgroups of whose centralizer orders are 12 or 6.

Direct computation on the 134 candidates shows that 5 groups with centralizer equal to 12 and 5 groups with centralizer order equal to 6 can occur as the Galois group of $f$ (note: there are 8 transitive subgroups of $S_{12}$ with centralizer order equal to 6, but only 5 have the correct filtration). We identify these groups in the table below using the transitive numbering system first introduced in [7]. We also give an alternative notation (in the second column), which is based on naming system currently implemented in [8].

| | |
|---|---|
| 12T1 | $C_{12}$ |
| 12T2 | $C_6 C_2$ |
| 12T3 | $D_6$ |
| 12T4 | $A_4$ |
| 12T5 | $1/2[3:2]4$ |
| 12T14 | $D_4 C_3$ |
| 12T15 | $1/2[3:2]dD(4)$ |
| 12T18 | $[3^2]E(4)$ |
| 12T19 | $[3^2]4$ |
| 12T42 | $C_6 \wr C_2$ |

## 7.4   Computation of Galois Groups

While most methods for the determination of Galois groups rely on the machinery of resolvent polynomials [10, 23, 24], ours does not. Instead, we use the list of the Galois groups of the Galois closures of the proper nontrivial subfields of the extension. We call this list the *subfield content* of $f$.

**Definition 2.** Let $f$ be an irreducible monic polynomial defining the extension $K/\mathbf{Q}_2$ with Galois group $G$. Suppose $K$ has $s$ proper nontrivial subfields up to isomorphism. Suppose these subfields have defining polynomials $f_1, \ldots, f_s$. Let $d_i$ denote the degree of $f_i$ and let $G_i$ be the Galois group of $f_i$ over $\mathbf{Q}_2$. Then $G_i$ is a transitive subgroup of $S_{d_i}$. Let $j_i$ denote the $T$-number of $G_i$ (as in [8]). The *subfield content* of $f$ is the set

$$\{d_1 \mathrm{T} j_1, d_2 \mathrm{T} j_2, \ldots, d_s \mathrm{T} j_s\},$$

customarily sorted in increasing order, first by $d_i$, then by $j_i$.

*Example 1.* For example, consider the first polynomial in Table 7.2, which defines the unique unramified degree 12 extension of $\mathbf{Q}_2$. Thus the Galois group $G$ of this polynomial is cyclic of order 12. Since the transitive group notation in [8] lists cyclic groups first, the $T$-number of $G$ is 12T1. By the fundamental theorem of Galois theory, since $G$ has a unique cyclic subgroup for every divisor of its order, $f$ has unique subfields of degrees 2, 3, 4, and 6. The Galois groups of these subfields are cyclic, and thus the subfield content of $f$ is {2T1, 3T1, 4T1, 6T1}.

*Example 2.* As another example, consider the 15th polynomial in Table 7.3, which is $f = x^{12} + 2x^6 + 4$. The stem field of $f$ clearly has subfields defined by the polynomials $x^6 + 2x^3 + 4$ and $x^4 + 2x^2 + 4$. Using [12], we see that the degree 6 polynomial has Galois group 6T5 $= C_3 \wr C_2$ and the degree 4 polynomial has 4T2 $= V_4$ as its Galois group. Since $V_4$ has three quadratic subfields, we know the subfield content of $f$ must contain the set {2T1, 2T1, 2T1, 4T2, 6T5}. Consulting Table 7.5, we see that this set must be equal to the subfield content of $f$, as no other option is possible. Notice this also proves that the Galois group of $f$ is 12T18.

In general, to compute the subfield content of one of our polynomials $f$, we can make use of the complete lists of quadratic, cubic, quartic, and sextic 2-adic fields determined in [12] (these lists include defining polynomials along with their Galois groups). For each polynomial in these lists, we can use Panayi's $p$-adic root-finding algorithm [16, 18] to test if the polynomial has a root in the field defined by $f$. If it does, then this polynomial defines a subfield of the field defined by $f$. Continuing in this way, it is straightforward to compute the subfield content of $f$.

We could also compute subfield content by realizing each degree 12 extension as a quadratic extension of a sextic 2-adic field. This approach can reduce the number of times Panayi's root-finding algorithm is used to compute the subfield content. Details of this approach can be found in [2].

The process of employing the subfield content of a polynomial to identify its Galois group is justified by the following result.

**Proposition 1.** *The subfield content of a polynomial is an invariant of its Galois group (thus it makes sense to speak of the subfield content of a transitive group).*

*Proof.* Suppose the polynomial $f$ defines an extension $L/K$ of fields, and let $G$ denote the Galois group of $f$. Let $E$ be the subgroup fixing $L/K$, arising from the Galois correspondence. The nonisomorphic subfields of $L/K$ correspond to the intermediate subgroups $F$, up to conjugation, such that $E \leq F \leq G$. Furthermore, if $K'$ is a subfield and $F$ is its corresponding intermediate group, then the Galois group of the normal closure of $K'$ is equal to the permutation representation of $G$ acting on the cosets of $F$ in $G$. Consequently, every polynomial with Galois group $G$ must have the same subfield content, and this quantity can be determined by a purely group-theoretic computation.                                                      □

Therefore, if we know that the Galois group of a polynomial $f$ must be contained in some set $S$ of transitive subgroups, and if the subfield contents for the groups in $S$

**Table 7.4** Subfield content for transitive subgroups of $S_{12}$ that have centralizer order 12

| T | Subfields | Polynomials |
|---|---|---|
| 12T1 | 2T1, 3T1, 4T1, 6T1 | 1, 3, 7, 8, 20, 21, 22, 23, 24, 25, 26, 27 |
| 12T2 | 2T1, 2T1, 2T1, 3T1, 4T2, 6T1, 6T1, 6T1 | 4, 5, 6, 15, 16, 17, 18 |
| 12T3 | 2T1, 2T1, 2T1, 3T2, 4T2, 6T2, 6T3, 6T3 | 9, 11, 13 |
| 12T4 | 3T1, 4T4, 6T4 | 19 |
| 12T5 | 2T1, 3T2, 4T1, 6T2 | 2, 10, 12, 14 |

The **Polynomials** column references row numbers in Table 7.2; the corresponding polynomials have the indicated Galois group

**Table 7.5** Subfield content for transitive subgroups of $S_{12}$ that have centralizer order 6

| T | Subfields | Polynomials |
|---|---|---|
| 12T14 | 2T1, 3T1, 4T3, 6T1 | 2, 3, 4, 5, 24, 25, 26, 27, 28, 29, 30, 31, 32<br>33, 34, 35, 36, 37, 38, 39, 40, 41, 42, 43, 44<br>45, 46, 47, 48, 49, 50, 51, 52, 53, 54, 55 |
| 12T15 | 2T1, 3T2, 4T3, 6T2 | 6, 11, 19, 21 |
| 12T18 | 2T1, 2T1, 2T1, 4T2, 6T5 | 7, 15, 16 |
| 12T19 | 2T1, 4T1, 6T5 | 1, 12, 17, 18 |
| 12T42 | 2T1, 4T3, 6T5 | 8, 9, 10, 13, 14, 20, 22, 23 |

The **Polynomials** column references row numbers in Table 7.5; the corresponding polynomials have the indicated Galois group

are all different, we can uniquely determine the Galois group of $f$ by computing its subfield content and matching it with its appropriate Galois group's subfield content.

In light of this observation, our approach for determining the Galois groups of the polynomials in Tables 7.2 and 7.3 involves three steps: (1) compute the subfield content for each of the possible ten Galois groups mentioned at the end of Sect. 7.3; (2) compute the subfield content for each of the 82 polynomials under consideration; (3) match up the polynomial's subfield content with the appropriate Galois group's subfield content to determine the Galois group of the polynomial.

Table 7.4 shows the subfield content for each transitive group of $S_{12}$ whose centralizer order is 12. The final column gives the row numbers of all polynomials in Table 7.2 that have the corresponding Galois group. Similarly, Table 7.5 shows the subfield content for each transitive subgroup of $S_{12}$ whose centralizer order is 6. The final column in this table references row numbers of polynomials in Table 7.3. In each table, the entries in column **Subfields** were computed with [8].

As a final note, we can compute subfield content for the remaining 124 transitive subgroups of $S_{12}$ that are possible Galois groups of degree 12 2-adic fields. Except for the unique group with centralizer order equal to 3 and a few groups with centralizer equal to 4, none of these groups can be distinguished solely by their subfield content. A complete description of subfield contents for the remaining 124 transitive groups of $S_{12}$ can be found in [2]. Identifying the Galois groups of the remaining 5,411 degree 12 2-adic fields from among these groups requires other methods and is the subject of ongoing research.

**Acknowledgments** The authors would like to thank the anonymous reviewer for his/her helpful comments. The authors would also like to thank Elon University for supporting this project through internal grants.

# References

1. Amano, S.: Eisenstein equations of degree $p$ in a p-adic field. J. Fac. Sci. Univ. Tokyo Sect. IA Math. **18**, 1–21 (1971). MR MR0308086 (46 #7201)
2. Awtrey, C.: Dodecic local fields. ProQuest LLC, Ann Arbor, MI. Ph.D. Thesis, Arizona State University (2010). MR 2736787
3. Awtrey, C.: On Galois groups of totally and tamely ramified sextic extensions of local fields. Int. J. Pure Appl. Math. **70**(6), 855–863 (2011)
4. Awtrey, C.: Dodecic 3-adic fields. Int. J. Number Theory **8**(4), 933–944 (2012). MR 2926553
5. Awtrey, C.: Masses, discriminants, and Galois groups of tame quartic and quintic extensions of local fields. Houston J. Math. **38**(2), 397–404 (2012). MR 2954644
6. Awtrey, C., Edwards, T.: Dihedral $p$-adic fields of prime degree. Int. J. Pure Appl. Math. **75**(2), 185–194 (2012)
7. Butler, G., McKay, J.: The transitive groups of degree up to eleven. Comm. Algebra **11**(8), 863–911 (1983). MR MR695893 (84f:20005)
8. GAP Group: The, GAP – Groups, Algorithms, Programming, Version 4.4.12. Available from http://www.gap-system.org/ (2008)
9. Greve, C., Pauli, S.: Ramification polygons, splitting fields, and Galois groups of Eisenstein polynomials. Int. J. Number Theory. **8**(6), 1401–1424 (2012). MR 2965757
10. Hulpke, A.: Techniques for the computation of Galois groups. In: Algorithmic Algebra and Number Theory (Heidelberg, 1997), pp. 65–77. Springer, Berlin (1999). MR 1672101 (2000d:12001)
11. Jones, J.W., Roberts, D.P.: Nonic 3-adic fields. In: Algorithmic Number Theory. Lecture Notes in Computer Science, vol. 3076, pp. 293–308. Springer, Berlin (2004). MR MR2137362 (2006a:11156)
12. Jones, J.W., Roberts, D.P.: A database of local fields. J. Symb. Comput. **41**(1), 80–97 (2006). MR 2194887 (2006k:11230)
13. Jones, J.W., Roberts, D.P.: Octic 2-adic fields. J. Number Theory **128**(6), 1410–1429 (2008). MR MR2419170 (2009d:11163)
14. Krasner, M.: Nombre des extensions d'un degré donné d'un corps p-adique, Les Tendances Géom. en Algèbre et Théorie des Nombres, pp. 143–169. Editions du Centre National de la Recherche Scientifique, Paris (1966). MR 0225756 (37 #1349)
15. Lang, S.: Algebraic Number Theory, 2nd edn. Graduate Texts in Mathematics, vol. 110, Springer, New York (1994). MR 1282723 (95f:11085)
16. Panayi, P.: Computation of leopoldt's $p$-adic regulator. Ph.D. Thesis, University of East Anglia (December 1995)
17. PARI Group: The, Bordeaux, PARI/GP, Version 2.5.3. Available from http://pari.math.u-bordeaux.fr/ (2012)
18. Pauli, S., Roblot, X.F.: On the computation of all extensions of a $p$-adic field of a given degree. Math. Comp. **70**(236), 1641–1659 (2001, electronic). MR 1836924 (2002e:11166)
19. Romano, D.S.: Galois groups of strongly Eisenstein polynomials. ProQuest LLC, Ann Arbor, MI. Ph.D. Thesis, University of California, Berkeley (2000). MR 2701040
20. Royle, G.F.: The transitive groups of degree twelve. J. Symb. Comput. **4**(2), 255–268 (1987). MR MR922391 (89b:20010)
21. Serre, J.P.: Une "formule de masse" pour les extensions totalement ramifiées de degré donné d'un corps local. C. R. Acad. Sci. Paris Sér. A-B **286**(22), A1031–A1036 (1978). MR 500361 (80a:12018)

22. Serre, J.P.: Local Fields. Graduate Texts in Mathematics, vol. 67, Springer, New York (1979). Translated from the French by Marvin Jay Greenberg. MR 554237 (82e:12016)
23. Soicher, L., McKay, J.: Computing Galois groups over the rationals. J. Number Theory **20**(3), 273–281 (1985). MR MR797178 (87a:12002)
24. Stauduhar, R.P.: The determination of Galois groups. Math. Comp. **27**, 981–996 (1973). MR 0327712 (48 #6054)

# Chapter 8
# Laplace Equations for Real Semisimple Associative Algebras of Dimension 2, 3 or 4.

James S. Cook, W. Spencer Leslie, Minh L. Nguyen, and Bailu Zhang

## 8.1 Introduction

In 1893 Scheffers [11] wrote a foundational paper on a theory of differentiation on a commutative unital algebra over $\mathbb{C}$. Then in 1900 Hausdorff [6] and 1933 Ringleb [9] extended the theory of analytic functions to noncommutative cases. In 1936 Spampinato [12] used the regular representation of the algebra to define differentiability for commutative algebras. In 1928 Ketchum [7] found results about power series of algebra variables. However, the background we present in this paper is most aligned with the results of Ward [15, 16] and Wagner [14].

The organization of this paper is as follows: in Sect. 8.2 we review the essentials of advanced calculus and associative algebras over $\mathbb{R}$. In particular, we develop three views of the algebra considered; $\mathscr{A} = \mathbb{R}^n$ the primary object, $L(\mathscr{A})$ the left-linear maps, and $M_{\mathscr{A}}$ the left regular representation. In Sect. 8.3 we explain how that the differential of an $\mathscr{A}$-differentiable function on $\mathscr{A}$ takes values in $L(\mathscr{A})$. We also observe that the Jacobian takes values in $M_{\mathscr{A}}$ and this requirement is equivalent to the generalized Cauchy Riemann equations. In Sect. 8.4 we turn to the question of generalizing Laplace's equation. We present an $n$-th order partial differential equation which we conjecture is solved by solutions of the generalized Cauchy Riemann equations for any semisimple associative algebra over $\mathbb{R}$. It should be mentioned that Wagner constructed Laplace equations for the special case of Frobenius algebras [14]. Our $\mathscr{A}$-Laplacian has the advantage of applying to noncommutative as well as commutative semisimple algebras.

J.S. Cook (✉) • W.S. Leslie • M.L. Nguyen • B. Zhang
Liberty University, 1971 University Blvd, Lynchburg, VA 24502, USA
e-mail: jcook4@liberty.edu; wsleslie@liberty.edu; mlnguyen@liberty.edu; bzhang2@liberty.edu

J. Rychtář et al. (eds.), *Topics from the 8th Annual UNCG Regional Mathematics and Statistics Conference*, Springer Proceedings in Mathematics & Statistics 64, DOI 10.1007/978-1-4614-9332-7_8, © Springer Science+Business Media New York 2013

## 8.2 Preliminaries

### 8.2.1 Differential Calculus on $\mathbb{R}^n$

The theory of differential calculus on $\mathbb{R}^n$ is the natural extension of calculus for functions on $\mathbb{R}$. Recall that $f : \mathbb{R} \to \mathbb{R}$ has a derivative $f'(a)$ at $x = a$ if

$$f'(a) = \lim_{h \to 0} \frac{f(a + h) - f(a)}{h}.$$

Alternatively, we can express the condition above as

$$\lim_{h \to 0} \frac{f(a + h) - f(a) - f'(a)h}{h} = 0.$$

This gives an implicit definition for $f'(a)$. This generalizes to higher dimensions as follows. For $F : U \subseteq \mathbb{R}^n \to \mathbb{R}^n$ if there exists a linear transformation $dF_a : \mathbb{R}^n \to \mathbb{R}^n$ such that

$$\lim_{h \to 0} \frac{F(a + h) - F(a) - dF_a(h)}{||h||} = 0,$$

where $||h||$ is the norm of a vector $h \in \mathbb{R}^n$, then we say that $F$ is differentiable at $a$ with **differential** $dF_a$. The matrix of the linear transformation $dF_a : \mathbb{R}^n \to \mathbb{R}^n$ is called the **Jacobian matrix** $F'(a) \in \mathbb{R}^{n \times n}$ or simply the **derivative** of $F$ at $a$. It follows that the components of the Jacobian matrix have the form $F'(a)_{ij} = \partial_j F_i(a)$ where $\partial_j$ denotes partial differentiation with respect to the $j$-th Cartesian coordinate. If the partial derivatives of $F_1, F_2, \ldots, F_n$ are continuous on $U$, then we say $F$ is continuously differentiable on $U$. A well-known theorem of advanced calculus states that continuous differentiability implies differentiability.

### 8.2.2 Associative Algebras on $\mathbb{R}^n$

To construct an algebra on $\mathbb{R}^n$ it suffices to define a multiplication on the standard basis $e_1, e_2, \ldots, e_n$. Denoting $\star : \mathbb{R}^n \times \mathbb{R}^n \to \mathbb{R}^n$ we need to supply constants $C_{ij}^k \in \mathbb{R}$ such that

$$e_i \star e_j = \sum_{k=1}^{n} C_{ij}^k e_k.$$

If $v, w \in \mathbb{R}^n$, then we define $v \star w$ by linearly extending the multiplication for the standard basis;

$$v \star w = \left( \sum_{i=1}^n v_i e_i \right) \star \left( \sum_{j=1}^n w_j e_j \right) = \sum_{i=1}^n \sum_{j=1}^n v_i w_j (e_i \star e_j) = \sum_{i,j,k=1}^n C_{ij}^k v_i w_j e_k.$$

The algebra is technically a pair $(\mathscr{A}, \star)$; however, we adopt the usual practice of refering to the pointset $\mathscr{A}$ as the algebra when the operation $\star$ is unambiguous. Also, for many standard examples we use juxtaposition rather than $\star$ to denote the product.

In the study of unital algebras it sometimes convenient[1] to set $e_1 = (1, 0, 0, \ldots, 0) = 1$ where 1 is the multiplicative unity in the algebra. Often in such discussions the structure constants are instead replaced by relations between generators which defined the algebra. As a simple principle, in such examples, it is understood that we multiply objects by the usual distributive rules paired with the given relation(s). For example, $i^2 = -1$ extended linearly defines the complex number system. Or $j^2 = 1$ extend linearly defines the hyperbolic number system.

We say $(\mathscr{A}, \star_\mathscr{A})$ and $(\mathscr{B}, \star_\mathscr{B})$ are isomorphic and write $\mathscr{A} \approx \mathscr{B}$ if and only if there exists a bijective linear transformation $\Phi : \mathscr{A} \to \mathscr{B}$ such that $\Phi(x \star_\mathscr{A} y) = \Phi(x) \star_\mathscr{B} \Phi(y)$ for all $x, y \in \mathscr{A}$. Furthermore, a commutative associative algebra $\mathscr{A}$ is called **semisimple** if its Jacobson radical is trivial.

The classification of associative, semisimple algebras over $\mathbb{R}$ was given by $E$. Cartan in 1884 [2]. See Chap. 2 of [1] for further historical and mathematical details. That said, we provide a classification argument based on several slightly more modern sources. In particular, recall that Frobenius Theorem [5] states that the only finite-dimensional division algebras over $\mathbb{R}$ are $\mathbb{R}$, $\mathbb{C}$ and the quaternions $\mathbb{H}$. Next, recall that Wedderburn's Theorem states that, up to isomorphism, any semisimple algebra over $\mathbb{R}$ is formed by direct sums of matrix algebras over the division rings of $\mathbb{R}$ (see pp. 855–856, Theorem 4 part (5) of Wedderburn's Theorem in [3]). Therefore, the only semisimple associative algebras over $\mathbb{R}$ are isomorphic to direct sums of the matrix algebras over $\mathbb{R}$, $\mathbb{C}$, and $\mathbb{H}$. We use $\mathbb{R}_m$, $\mathbb{C}_m$, and $\mathbb{H}_m$ to denote the representation of the $m \times m$ matrix algebras on $\mathbb{R}^n$ with $n = m^2, 4m^2$, and $16m^2$, respectively. Our focus in this article concerns semisimple associative algebras of dimension $1, 2, 3, 4$ hence the only nontrivial matrix algebra we consider is that of $\mathbb{R}^{2\times2} \approx \mathbb{R}_2$. Consider, if $n = 2$, the theory allows only two semisimple algebras up to isomorphism; namely $\mathbb{R} \oplus \mathbb{R}$ and $\mathbb{C}$. If $n = 3$, we have semisimple algebras $\mathbb{R} \oplus \mathbb{R} \oplus \mathbb{R}$ and $\mathbb{R} \oplus \mathbb{C}$. In dimension $n = 4$ we have commutative and noncommutative examples. For $n = 4$ commutative, $\mathbb{R} \oplus \mathbb{R} \oplus \mathbb{R} \oplus \mathbb{R}$, $\mathbb{R} \oplus \mathbb{R} \oplus \mathbb{C}$, and $\mathbb{C} \oplus \mathbb{C}$. For $n = 4$ noncommutative we have quaternions $\mathbb{H}$ and $\mathbb{R}_2$.

---

[1]This is not always assumed in this article.

## 8.2.3   Left Regular Representations

Suppose $\mathscr{A} = \mathbb{R}^n$ is an $n$-dimensional unital associative algebra over $\mathbb{R}$ with multiplication denoted by $\star$. A linear mapping $T : \mathbb{R}^n \to \mathbb{R}^n$ is **left-$\mathscr{A}$-linear** if and only if $T(x \star y) = T(x) \star y$ for all $x, y \in \mathscr{A}$. Note that a left-linear map is uniquely defined by its value on the unity:

$$T(x) = T(1 \star x) = T(1) \star x.$$

This means $T$ is a **left-multiplication** map of the algebra $\mathscr{A}$. We define $L_v(x) = v \star x$ and observe that an arbitrary left-multiplication map $L_v : \mathscr{A} \to \mathscr{A}$ is a linear transformation which, by associativity[2] is left-linear:

$$L_v(x \star y) = v \star (x \star y) = (v \star x) \star y = L_v(x) \star y.$$

Therefore, we can identify the set of left-multiplication maps and the set of left-linear maps as the same set of mappings on a unital associative algebra.

**Definition 1.**  Let $\mathscr{A}$ be a unital associative algebra over $\mathbb{R}$ then we denote the set of **left-linear** maps by $L(\mathscr{A})$.

Moreover, as $L_{x \star y} = L_x \circ L_y$ for all $x, y \in \mathscr{A}$ we find $L(\mathscr{A})$ forms a subalgebra of the endomorphisms of $\mathscr{A}$ which is isomorphic to $\mathscr{A}$. To make this isomorphism explicit it helps to develop some notation. Recall the standard matrix of $T$ is given by: $[T] = [T(e_1)|T(e_2)|\cdots|T(e_n)]$. However, $e_j = 1 \star e_j$ hence $T(e_j) = T(1 \star e_j) = T(1) \star e_j$. Consequently:

$$[T] = [T(1) \star e_1|T(1) \star e_2|\cdots|T(1) \star e_n].$$

Let $t_1, t_2, \ldots t_n \in \mathbb{R}$ are given such that $T(1) = t_1 e_1 + t_2 e_2 + \cdots + t_n e_n$. The matrix $[T]$ is uniquely specified by the constants $t_i$ and the structure constants of the multiplication. In particular since $e_i \star e_j = \sum_{k=1}^n C_{ij}^k e_k$ we find:

$$T(e_j) = T(1) \star e_j = \sum_{i=1}^n t_i e_i \star e_j = \sum_{i,k=1}^n t_i C_{ij}^k e_k \;\; \Rightarrow \;\; [T]_{kj} = \sum_{i,k=1}^n t_i C_{ij}^k.$$

Therefore, $\Phi : \mathscr{A} \to L(\mathscr{A})$ with $[\Phi(t)]_{kj} = \sum_{i,k=1}^n t_i C_{ij}^k$ gives the isomorphism $\mathscr{A} \approx L(\mathscr{A})$.

Furthermore, the correspondence of each left-multiplication map to its standard matrix provides an isomorphic image of $L(\mathscr{A})$ in $\mathbb{R}^{n \times n}$.

---

[2]To generalize to nonassociative algebras we would need a different technique, see, for example, the paper on Cayley-Dickson calculus [8] which makes due with the weaker property of power associativity.

**Definition 2.** Let $M_{\mathscr{A}} = \{A \in \mathbb{R}^{n \times n} \mid A = [T] \; for \; some \; T \in L(\mathscr{A})\}$. We say the $n \times n$ matrix which corresponds to $v \in \mathscr{A}$ is the **left regular representation** of $v$.

Note $M_{\mathscr{A}}$ forms a subalgebra of $\mathbb{R}^{n \times n}$ with respect to matrix multiplication. We have three representations of the algebra considered: the pointset $\mathscr{A} = \mathbb{R}^n$ which we take as primary, the set of left-$\mathscr{A}$-linear maps $L(\mathscr{A})$ and perhaps most interestingly the left regular representation $M_{\mathscr{A}}$. For convenience to the reader and clarity of exposition we now list the left regular representations for our list of examples.

*Example 1.* The **real numbers** with their usual addition and multiplication is an associative algebra over $\mathbb{R}$. If $a \in \mathbb{R}$, then $[a] \in M_{\mathbb{R}} = \mathbb{R}^{1 \times 1}$ is its left regular representation. Usually we will not distinguish between $a$ and $[a]$.

Two-dimensional examples are a bit more exciting. Let it be noted that for the next three examples we use the notation $1 = e_1 = (1,0)$ and $e_2$ is assigned to be the generator of the algebra. We use juxtaposition rather than $\star$ in the interest of matching the standard literature.

*Example 2.* The **complex numbers** are defined by $\mathbb{C} = \mathbb{R} \oplus i\mathbb{R}$ where $i^2 = -1$. If $a + ib, c + id \in \mathbb{C}$, then $(a + ib)(c + id) = ac + iad + ibc + i^2bd = ac - bd + i(ad + bc)$. Note every nonzero complex number $a + ib$ has multiplicative inverse $\frac{a - ib}{a^2 + b^2}$, thus $\mathbb{C}$ has no zero-divisors. Note $A = \begin{bmatrix} a & -b \\ b & a \end{bmatrix} \in M_{\mathbb{C}}$ represents $a + ib$.

*Example 3.* The **hyperbolic** numbers are given by $\mathscr{H} = \mathbb{R} \oplus j\mathbb{R}$ where $j^2 = 1$. If $a + jb, c + jd \in \mathscr{H}$, then $(a + jb)(c + jd) = ac + adj + jbc + j^2bd = ac + bd + j(ad + bc)$. There are zero-divisors in $\mathscr{H}$. Observe $(a + ja)(a - ja) = 0$ for $a \neq 0$. Observe $A = \begin{bmatrix} a & b \\ b & a \end{bmatrix} \in M_{\mathscr{H}}$ represents $a + jb$.

*Example 4.* The **dual numbers** are given by $\mathscr{N} = \mathbb{R} \oplus \eta\mathbb{R}$ where $\eta^2 = 0$. If $a + \eta b, c + \eta d \in \mathscr{N}$, then

$$(a + \eta b)(c + \eta d) = ac + ad\eta + bc\eta + \eta^2 bd = ac + (ad + bc)\eta.$$

Observe $\mathscr{N}$ has many zero divisors. The dual number $a + \eta b$ has representative $A = \begin{bmatrix} a & 0 \\ b & a \end{bmatrix} \in M_{\mathscr{N}}$. Observe that the ideal generated by $\eta$ is nontrivial and is found in the Jacobson radical; hence, the dual numbers $\mathscr{N}$ are not semisimple (see example 3.5.6 on p. 47 of [4] for a related discussion).

We now turn to higher-dimensional associative algebras. There are additional non-semisimple examples for dimensions 3 and 4, but our focus is on the semisimple case.

*Example 5.* Let $\mathscr{A} = \mathbb{R} \oplus j\mathbb{R} \oplus j^2\mathbb{R}$ where $j^3 = 1$. The matrix representatives of these numbers have an interesting pattern; note: $A \in \mathrm{M}_{\mathscr{A}}$ implies $A = \begin{bmatrix} a & c & b \\ b & a & c \\ c & b & a \end{bmatrix}$.
We note an isomorphism $\mathscr{A} \approx \mathbb{R} \times \mathbb{C}$ is given by mapping $j$ to $(1, \omega)$ where $\omega$ is a third root of unity.

*Example 6.* Let $\mathscr{A} = \mathbb{R} \times \mathscr{H}$ where $1 = (1, 1 + 0j)$. In the natural basis this gives representatives $A \in \mathrm{M}_{\mathscr{A}}$ which are block-diagonal; $A = \begin{bmatrix} a & 0 & 0 \\ 0 & b & c \\ 0 & c & b \end{bmatrix}$. We can show this algebra is isomorphic to $\mathbb{R} \times \mathbb{R} \times \mathbb{R}$ with the Hadamard product $(a_1, a_2, a_3) \star (b_1, b_2, b_3) = (a_1 b_1, a_2 b_2, a_3 b_3)$.

*Example 7.* Let $\mathscr{A} = \mathbb{R} \oplus j\mathbb{R} \oplus j^2\mathbb{R} \oplus j^3\mathbb{R}$ where $j^4 = 1$. Much as was the pattern for the $j^2 = 1$ (hyperbolic numbers) or $j^3 = 1$ we find a beautiful pattern: $A \in \mathrm{M}_{\mathscr{A}}$ implies $A = \begin{bmatrix} a & d & c & b \\ b & a & d & c \\ c & b & a & d \\ d & c & b & a \end{bmatrix}$. This algebra is naturally isomorphic to $\mathbb{C} \oplus \mathscr{H}$
which is clearly isomorphic to $\mathbb{C} \oplus \mathbb{R} \oplus \mathbb{R}$.

*Example 8.* Let $\mathscr{A} = \mathscr{H} \times \mathscr{H}$ where $1 = (1 + 0j, 1 + 0j)$. This means $(1, 1)$ is naturally represented by the identity matrix. In total we have once more a block-diagonal representation: $A \in \mathrm{M}_{\mathscr{A}}$ implies $A = \left[ \begin{array}{cc|cc} a & b & 0 & 0 \\ b & a & 0 & 0 \\ \hline 0 & 0 & c & d \\ 0 & 0 & d & c \end{array} \right]$ and this
matrix represents $(a + bj, c + dj)$. We can show this algebra is isomorphic to $\mathbb{R} \times \mathbb{R} \times \mathbb{R} \times \mathbb{R}$ with the Hadamard product $(a_1, a_2, a_3, a_4) * (b_1, b_2, b_3, b_4) = (a_1 b_1, a_2 b_2, a_3 b_3, a_4 b_4)$.

*Example 9.* Let $\mathscr{A} = \mathbb{C} \times \mathbb{C}$. Here we study the problem of two complex variables. In this algebra $(1 + 0i, 1 + 0i)$ corresponds to the identity and hence $(1, 1)$ is naturally represented by the identity matrix. In total we have once more a block-diagonal representation: $A \in \mathrm{M}_{\mathscr{A}}$ implies $A = \left[ \begin{array}{cc|cc} a & -b & 0 & 0 \\ b & a & 0 & 0 \\ \hline 0 & 0 & c & -d \\ 0 & 0 & d & c \end{array} \right]$ and this matrix represents
$(a + bi, c + di)$.

*Example 10.* Let $\mathbb{H} = \mathbb{R} \oplus i\mathbb{R} \oplus j\mathbb{R} \oplus k\mathbb{R}$ where $i^2 = j^2 = k^2 = -1$ and $ij = k$. These are Hamilton's famed **quaternions**. We can show $ij = -ji$; hence, these are not commutative. With respect to the natural basis $e_1 = 1, e_2 = i, e_3 = j, e_4 = k$ we find the matrix representative of $a + ib + cj + dk$ is as follows:

$$A = \begin{bmatrix} a & -b & -c & -d \\ b & a & -d & c \\ c & d & a & -b \\ d & -c & b & a \end{bmatrix} \in M_{\mathbb{H}}.$$

*Example 11.* Let $\mathscr{A} = \mathbb{R}_2$ with the multiplication $\star$ induced from the multiplication of $2 \times 2$ matrices. This again forms a noncommutative algebra. In particular, this multiplication is induced in the natural manner:

$$\begin{bmatrix} a & b \\ c & d \end{bmatrix} \begin{bmatrix} t & x \\ y & z \end{bmatrix} = \begin{bmatrix} at + by & ax + bz \\ ct + dy & cx + dz \end{bmatrix}.$$

It follows that $(a,b,c,d) \star (t,x,y,z) = (at + by, ax + bz, ct + dy, cx + dz)$. We can read from this multiplication that the representative of $(a,b,c,d) \in \mathbb{R}_2$ is given by

$$A = \begin{bmatrix} a & 0 & b & 0 \\ 0 & a & 0 & b \\ c & 0 & d & 0 \\ 0 & c & 0 & d \end{bmatrix} = \begin{bmatrix} aI & bI \\ cI & dI \end{bmatrix} \in M_{\mathscr{A}}.$$

*Remark 1.* The examples above provide a particular representation of each algebra. There are other ways to place each of these algebras on $\mathbb{R}^n$. For example, the Hadamard product gives diagonal left representations $M_{\mathscr{A}}$. We choose to study an isomorphic product which provides a less sparse regular representation. The interplay between isomorphisms of algebras and coordinate changes of partial differential equations is one of the features which captures our interest in this problem. Note that [14, 16] show that $\mathscr{A}$-differentiability is preserved under an algebra isomorphism. See p. 457 of [14], but note that we prefer to replace the term *analytic* with $\mathscr{A}$-differentiable.

## 8.3   Differential Calculus on an Associative Algebra

Suppose $(\mathscr{A}, \star)$ is an associative unital algebra over $\mathbb{R}$ where as a pointset $\mathscr{A} = \mathbb{R}^n$.

**Definition 3.** Let $U \subseteq \mathscr{A}$ and consider $f : U \to \mathscr{A}$. We say $f$ is $\mathscr{A}$-**differentiable** at $p \in U$ if and only if $f$ is differentiable at $p$ and the differential $d_p f \in L(\mathscr{A})$.

Since $d_p f \in L(\mathscr{A})$ we find $[d_p f] = f'(p) \in M_{\mathscr{A}}$. This means that $\mathscr{A}$-differentiability of $f$ implies the Jacobian matrix of $f$ is a left regular representation of $\mathscr{A}$. The statement $f'(p) \in M_{\mathscr{A}}$ implies equations amongst the partial derivatives of $f$ which are known as the **generalized Cauchy Riemann equations**.

*Remark 2.* It is interesting to note that Ward showed in [15] that if one is given a set of partial differential equations of a certain form, then it is possible to find an algebra for which the given equations form the generalized Cauchy Riemann equations.

From our discussion in 8.2.3 we can expect $n^2 - n$ generalized Cauchy Riemann equations. Since there is no danger of confusion we will simply refer to these as the *Cauchy Riemann* equations in what follows.

**Theorem 1.** *If $f, g$ are $\mathscr{A}$-differentiable at $p \in \mathscr{A}$, then $f + g$ and $cf$ are $\mathscr{A}$ differentiable for each $c \in \mathbb{R}$. Moreover, if $g$ is $\mathscr{A}$-differentiable at $f(p)$ and $f$ is $\mathscr{A}$-differentiable at $p \in \mathscr{A}$, then $g \circ f$ is $\mathscr{A}$-differentiable at $p$.*

*Proof.* From advanced calculus we know $d_p(f + g) = d_p f + d_p g$ and $d_p(cf) = c d_p f$ for all $c \in \mathbb{R}$. Furthermore, the chain-rule can be stated as $d_p(g \circ f) = d_{f(p)} g \circ d_p f$. We need to only show that $\mathscr{A}$-linearity is preserved in view of these formulas. Observe:

$$d_p(f + g)(v \star w) = d_p f(v \star w) + d_p g(v \star w)$$
$$= d_p f(v) \star w + d_p g(v) \star w$$
$$= [d_p f(v) + d_p g(v)] \star w$$
$$= d_p(f + g)(v) \star w.$$

Thus $d_p(f + g) \in L(\mathscr{A})$. The proof that $c d_p f \in L(\mathscr{A})$ is similar. Finally:

$$d_p(g \circ f)(v \star w) = d_{f(p)} g(d_p f(v \star w)) = d_{f(p)} g(d_p f(v) \star w) = d_{f(p)} g(d_p f(v)) \star w,$$

hence $d_p(g \circ f) \in L(\mathscr{A})$.                                                                          $\square$

The product of two functions on $\mathscr{A}$ is defined by $(f \star g)(p) = f(p) \star g(p)$.

**Theorem 2.** *If $f, g$ are $\mathscr{A}$-differentiable at $p \in \mathscr{A}$, then $d_p(f \star g)(v) = d_p f(v) \star g(p) + f(p) \star d_p g(v)$ for all $v \in \mathscr{A}$. However, $f \star g$ need not be $\mathscr{A}$-differentiable.*

*Proof.* The proof follows from direct calculation with the structure constants and the usual product rules for functions of $n$-real variables. Let $f = \sum_i f_i e_i$ and $g = \sum_j g_j e_j$ we calculate from $e_i \star e_j = \sum_k C_{ij}^k e_k$ that $f \star g = \sum_{i,j,k} f_i g_j C_{ij}^k e_k$. Observe:

$$\partial_l(f \star g) = \sum_{i,j,k} \partial_l(f_i g_j) C_{ij}^k e_k = \sum_{i,j,k} [(\partial_l f_i) g_j + f_i(\partial_l g_j)] C_{ij}^k e_k$$
$$= \sum_{i,j,k} (\partial_l f_i) g_j C_{ij}^k e_k + \sum_{i,j,k} f_i(\partial_l g_j) C_{ij}^k e_k$$
$$= \partial_l f \star g + f \star \partial_l g.$$

It follows that $d_p(f \star g)(v) = d_p f(v) \star g(p) + f(p) \star d_p g(v)$. Observe that $(d_p(f \star g))(v \star w) \neq (d_p(f \star g)(v)) \star w$ due to non-commutative examples. There is no general reason to allow the $\mathscr{A}$-element $w$ to commute past $g(p)$ without introducing unwanted terms. It follows that $f \star g$ is not generally $\mathscr{A}$-differentiable.

$\square$

*Remark 3.* $\mathscr{A}$-differentiability is a strong condition for noncommutative examples. In [10] Rosenfeld indicates that the only left and right differentiable functions in the noncommutative case are linear functions. We expect his claim applies in our context. The only $\mathscr{A}$-differentiable functions for noncommutative associative algebras are linear functions. On the other hand, the theorem above clearly suggests that polynomials of $\mathscr{A}$-variables will form $\mathscr{A}$-differentiable functions for commutative algebras.

We now turn to the explicit calculation of the Cauchy Riemann equations for our set of examples.

*Example 12.* If $f : U \subseteq \mathbb{R} \rightarrow \mathbb{R}$ is differentiable at $p \in \mathbb{R}$, then it follows $d_p f(h) = f'(p)h$ hence differentiability at $p$ implies $\mathbb{R}$-linearity of the differential. In other words, differentiability at $p \in \mathbb{R}$ implies $\mathbb{R}$-differentiability at $p$.

If $\mathscr{A} = \mathbb{R}^2$ then it is convenient to denote $f : \mathscr{A} \rightarrow \mathscr{A}$ by $f = ue_1 + ve_2$ where $u$ and $v$ are the component functions with respect to the basis $e_1$ and $e_2$. The Jacobian for a real-differentiable function is simply $f' = \begin{bmatrix} u_x & u_y \\ v_x & v_y \end{bmatrix}$. If we impose $f' \in M_{\mathscr{A}}$, then we must find certain relations on the components of the Jacobian, these are Cauchy Riemann equations.

*Example 13.* The standard Cauchy Riemann equations for $f = u + iv : \mathbb{C} \rightarrow \mathbb{C}$ are derived from Example 2. Following the pattern we find $f' = \begin{bmatrix} u_x & -v_x \\ v_x & u_x \end{bmatrix} = \begin{bmatrix} u_x & u_y \\ v_x & v_y \end{bmatrix}$ hence $u_x = v_y$ and $u_y = -v_x$.

*Example 14.* To determine the Cauchy Riemann equations for $f = u + jv : \mathscr{H} \rightarrow \mathscr{H}$ we use results derived in Example 3. By supposing the Jacobian matrix is a representative of the algebra $\mathscr{H}$ we find $f' = \begin{bmatrix} u_x & v_x \\ v_x & u_x \end{bmatrix} = \begin{bmatrix} u_x & u_y \\ v_x & v_y \end{bmatrix}$, hence $u_x = v_y$ and $u_y = v_x$.

*Example 15.* The Cauchy Riemann equations for $f = u + \eta v : \mathscr{N} \rightarrow \mathscr{N}$ are derived from Example 4. Imposing the pattern we find $f' = \begin{bmatrix} u_x & 0 \\ v_x & u_x \end{bmatrix} = \begin{bmatrix} u_x & u_y \\ v_x & v_y \end{bmatrix}$, hence $u_x = v_y$ and $u_y = 0$.

For three-dimensional examples it is convenient to denote $f = ue_1 + ve_2 + we_3$, hence the Jacobian for a real-differentiable function is simply $f' = \begin{bmatrix} u_x & u_y & u_z \\ v_x & v_y & v_z \\ w_x & w_y & w_z \end{bmatrix}$.

*Example 16.* Cauchy Riemann equations for $\mathscr{A} = \mathbb{R} \oplus j\mathbb{R} \oplus j^2\mathbb{R}$ are easily extrapolated from Example 5. Observe $f' = \begin{bmatrix} u_x & w_x & v_x \\ v_x & u_x & w_x \\ w_x & v_x & u_x \end{bmatrix} = \begin{bmatrix} u_x & u_y & u_z \\ v_x & v_y & v_z \\ w_x & w_y & w_z \end{bmatrix}$.

Therefore, the Cauchy Riemann equations are:

$$u_x = v_y = w_z, \qquad u_y = v_z = w_x, \qquad u_z = v_x = w_y.$$

*Example 17.* Cauchy Riemann equations for $\mathscr{A} = \mathbb{R} \times \mathscr{H}$ are easily lifted from Example 6. Observe $f' = \begin{bmatrix} u_x & 0 & 0 \\ 0 & v_y & w_z \\ 0 & w_z & v_y \end{bmatrix} = \begin{bmatrix} u_x & u_y & u_z \\ v_x & v_y & v_z \\ w_x & w_y & w_z \end{bmatrix}$. Therefore, the Cauchy Riemann equations are:

$$v_x = w_x = u_y = u_z = 0, \qquad w_z = v_y, \qquad w_y = v_z.$$

For the four-dimensional examples it is convenient to denote $f = \phi e_1 + ue_2 + ve_3 + we_4$ and we take Cartesian coordinates $(t, x, y, z)$ by default. It follows the Jacobian for a real-differentiable function is simply $f' = \begin{bmatrix} \phi_t & \phi_x & \phi_y & \phi_z \\ u_t & u_x & u_y & u_z \\ v_t & v_x & v_y & v_z \\ w_t & w_x & w_y & w_z \end{bmatrix}$.

*Example 18.* Cauchy Riemann equations for $\mathscr{A} = \mathbb{R} \oplus j\mathbb{R} \oplus j^2\mathbb{R} \oplus j^3\mathbb{R}$ are found from Example 7. Set $f' = \begin{bmatrix} \phi_t & \phi_x & \phi_y & \phi_z \\ \phi_z & \phi_t & \phi_x & \phi_y \\ \phi_y & \phi_z & \phi_t & \phi_x \\ \phi_x & \phi_y & \phi_z & \phi_t \end{bmatrix} = \begin{bmatrix} \phi_t & \phi_x & \phi_y & \phi_z \\ u_t & u_x & u_y & u_z \\ v_t & v_x & v_y & v_z \\ w_t & w_x & w_y & w_z \end{bmatrix}$. We find Cauchy Riemann equations:

$$\phi_t = u_x = v_y = w_z, \quad \phi_x = u_y = v_z = w_t,$$

and

$$\phi_y = u_z = v_t = w_x, \quad \phi_z = u_t = v_x = w_y.$$

*Example 19.* Cauchy Riemann equations for $\mathscr{A} = \mathscr{H} \times \mathscr{H}$ are found from

Example 8. Set $f' = \begin{bmatrix} \phi_t & \phi_x & 0 & 0 \\ \phi_x & \phi_t & 0 & 0 \\ 0 & 0 & v_y & v_z \\ 0 & 0 & v_z & v_y \end{bmatrix} = \begin{bmatrix} \phi_t & \phi_x & \phi_y & \phi_z \\ u_t & u_x & u_y & u_z \\ v_t & v_x & v_y & v_z \\ w_t & w_x & w_y & w_z \end{bmatrix}$. The Cauchy Riemann

equations are:

$$\phi_y = \phi_z = u_y = u_z = v_t = v_x = w_t = w_x = 0,$$

and

$$\phi_t = u_x, \quad \phi_x = u_t, \quad v_y = w_z, \quad v_z = w_y.$$

*Example 20.* Cauchy Riemann equations for $\mathscr{A} = \mathbb{C} \times \mathbb{C}$ are found from

Example 9. Set $f' = \begin{bmatrix} \phi_t & \phi_x & 0 & 0 \\ -\phi_x & \phi_t & 0 & 0 \\ 0 & 0 & v_y & v_z \\ 0 & 0 & -v_z & v_y \end{bmatrix} = \begin{bmatrix} \phi_t & \phi_x & \phi_y & \phi_z \\ u_t & u_x & u_y & u_z \\ v_t & v_x & v_y & v_z \\ w_t & w_x & w_y & w_z \end{bmatrix}$. The Cauchy

Riemann equations are:

$$\phi_y = \phi_z = u_y = u_z = v_t = v_x = w_t = w_x = 0,$$

and

$$\phi_t = u_x, \quad \phi_x = -u_t, \quad v_y = w_z, \quad v_z = -w_y.$$

*Example 21.* Cauchy Riemann equations for $\mathbb{H} = \mathbb{R} \oplus i\mathbb{R} \oplus j\mathbb{R} \oplus k\mathbb{R}$ are found from

Example 10. Set $f' = \begin{bmatrix} \phi_t & -u_t & -v_t & -w_t \\ u_t & \phi_t & -w_t & v_t \\ v_t & w_t & \phi_t & -u_t \\ w_t & -v_t & u_t & \phi_t \end{bmatrix} = \begin{bmatrix} \phi_t & \phi_x & \phi_y & \phi_z \\ u_t & u_x & u_y & u_z \\ v_t & v_x & v_y & v_z \\ w_t & w_x & w_y & w_z \end{bmatrix}$. The Cauchy

Riemann equations are:

$$\phi_t = u_x = v_y = w_z, \qquad u_t = -\phi_x = w_y = -v_z,$$

and

$$v_t = -w_x = -\phi_y = u_z, \qquad w_t = v_x = -u_y = -\phi_z.$$

*Example 22.* Cauchy Riemann equations for the matrix multiplication algebra of real $2 \times 2$ matrices are found from the left regular representations of $\mathscr{A} = \mathbb{R}_2$

given in Example 11. Set $f' = \begin{bmatrix} \phi_t & 0 & -\phi_y & 0 \\ 0 & \phi_t & 0 & \phi_y \\ v_t & 0 & v_y & 0 \\ 0 & v_t & 0 & v_y \end{bmatrix} = \begin{bmatrix} \phi_t & \phi_x & \phi_y & \phi_z \\ u_t & u_x & u_y & u_z \\ v_t & v_x & v_y & v_z \\ w_t & w_x & w_y & w_z \end{bmatrix}$. The Cauchy

Riemann equations are:

$$u_t = \phi_x = u_y = \phi_z = w_t = v_x = w_y = v_z = 0,$$

and

$$\phi_t = u_x, \quad \phi_y = u_z, \quad v_t = w_x, \quad v_y = w_z.$$

## 8.4   The $\mathscr{A}$-Laplacian

Our generalization of Laplace's equation for an associative unital algebra is intended to satisfy the following two criteria:

1. each component of an $\mathscr{A}$-differentiable function should solve the $\mathscr{A}$-Laplace equation,
2. the $\mathscr{A}$-Laplace equation is a single real partial differential equation.

Let $\Psi : \mathscr{A} \to M_{\mathscr{A}}$ to be the natural isomorphism described in Sect. 8.2.3. If $e_1, e_2, \ldots, e_n$ forms the standard basis for $\mathscr{A} = \mathbb{R}^n$, then let $E_j = \Psi(e_j)$, hence $E_1, E_2, \ldots, E_n$ forms a basis for $M(\mathscr{A})$. Hence define:

$$\triangle_{\mathscr{A}} = \det(E_1 \partial_1 + E_2 \partial_2 + \cdots + E_n \partial_n).$$

Formally this amounts to taking the determinant of the left regular representation of $(\partial_1, \partial_2, \ldots, \partial_n)$.

*Conjecture 1.* Suppose $\mathscr{A}$ is an associative, unital, semisimple algebra over $\mathbb{R}$ of dimension greater than 1. If $f = (u_1, u_2, \ldots, u_n)$ is $\mathscr{A}$-differentiable on $U \subseteq \mathscr{A}$, then each of the component functions $u_j$ solves $\triangle_{\mathscr{A}} u_j = 0$ on $U$.

In each of the semisimple examples with $2 \leq dim(\mathscr{A}) \leq 4$ we have checked by explicit computation that the solution set of the Cauchy Riemann equations is likewise in the solution set of the $\mathscr{A}$-Laplacian equation $\triangle_{\mathscr{A}} u = 0$. Example 23 explains why we must rule out $\mathscr{A} = \mathbb{R}$ and the necessity of semisimplicity is made manifest in Example 26.

### 8.4.1 The $\mathscr{A}$-Laplace Equations

*Example 23.* Consider $\mathscr{A} = \mathbb{R}$. In this case $e_1 = 1$ and $E_1 = [1]$, hence $\triangle_{\mathscr{A}} = det(\partial_x)$. The $\mathbb{R}$-Laplace equation is simply $\partial_x f = f'(x) = 0$. The only solutions in this one-dimensional case are constants. However, $\mathbb{R}$-differentiable functions include nonconstant examples. Consequently, the conjecture must begin at $n = 2$.

*Example 24.* For $\mathbb{C}$ we have $E_1 = \begin{bmatrix} 1 & 0 \\ 0 & 1 \end{bmatrix}$ and $E_2 = \begin{bmatrix} 0 & -1 \\ 1 & 0 \end{bmatrix}$, thus

$$\triangle_{\mathbb{C}} = det(E_1 \partial_x + E_2 \partial_y) = det \begin{bmatrix} \partial_x & -\partial_y \\ \partial_y & \partial_x \end{bmatrix} = \partial_x^2 + \partial_y^2.$$

We recognize $\triangle_{\mathbb{C}} u = u_{xx} + u_{yy} = 0$ as the standard Laplace equation of complex analysis.

*Example 25.* For $\mathscr{H}$ we have $E_1 = \begin{bmatrix} 1 & 0 \\ 0 & 1 \end{bmatrix}$ and $E_2 = \begin{bmatrix} 0 & 1 \\ 1 & 0 \end{bmatrix}$, thus

$$\triangle_{\mathscr{H}} = det(E_1 \partial_x + E_2 \partial_y) = det \begin{bmatrix} \partial_x & \partial_y \\ \partial_y & \partial_x \end{bmatrix} = \partial_x^2 - \partial_y^2.$$

Observe $\triangle_{\mathscr{H}} u = u_{xx} - u_{yy} = 0$ is the one-dimensional wave equation; it is the fundamental hyperbolic partial differential equation.

*Example 26.* For $\mathscr{N}$ we have $E_1 = \begin{bmatrix} 1 & 0 \\ 0 & 1 \end{bmatrix}$ and $E_2 = \begin{bmatrix} 0 & 0 \\ 1 & 0 \end{bmatrix}$, thus

$$\triangle_{\mathscr{N}} = det(E_1 \partial_x + E_2 \partial_y) = det \begin{bmatrix} \partial_x & 0 \\ \partial_y & \partial_x \end{bmatrix} = \partial_x^2.$$

Recall $f = u + \eta v$ is $\mathscr{N}$-differentiable if and only if $f$ is real differentiable and $u_x = v_y$ and $u_y = 0$. Notice that $v_x$ is free in this example. Let $f(x, y) = g(x)\eta$ then $u = 0$ and $v = g$ clearly satisfies $u_x = v_y$ and $u_y = 0$ however $\triangle_{\mathscr{N}} g = \partial_x^2 g$ which need not be zero. In this nonsemisimple case we see that $\mathscr{A}$-differentiability does not imply the $\mathscr{A}$-Laplace equation.

We have observed similar difficulty in other nonsemisimple examples which we do not present in this current report.

*Example 27.* Following Examples 5 and 16 if $\mathscr{A} = \mathbb{R} \oplus j\mathbb{R} \oplus j^2\mathbb{R}$, then

$$\Delta_{\mathscr{A}} = \det \begin{bmatrix} \partial_x & \partial_z & \partial_y \\ \partial_y & \partial_x & \partial_z \\ \partial_z & \partial_y & \partial_x \end{bmatrix} = \partial_x^3 + \partial_y^3 + \partial_z^3 - 3\partial_x\partial_y\partial_z.$$

*Example 28.* Following Examples 6 and 17 if $\mathscr{A} = \mathbb{R} \times \mathscr{H}$, then

$$\Delta_{\mathscr{A}} = \det \begin{bmatrix} \partial_x & 0 & 0 \\ 0 & \partial_y & \partial_z \\ 0 & \partial_y & \partial_z \end{bmatrix} = \partial_x(\partial_y^2 - \partial_z^2).$$

*Example 29.* Following Examples 7 and 18 if $\mathscr{A} = \mathbb{R} \oplus j\mathbb{R} \oplus j^2\mathbb{R} \oplus j^3\mathbb{R}$, then

$$\Delta_{\mathscr{A}} = \det \begin{bmatrix} \partial_t & \partial_z & \partial_y & \partial_x \\ \partial_x & \partial_t & \partial_z & \partial_y \\ \partial_y & \partial_x & \partial_t & \partial_z \\ \partial_z & \partial_y & \partial_x & \partial_t \end{bmatrix},$$

hence

$$\Delta_{\mathscr{A}} = \partial_t^4 - \partial_x^4 + \partial_y^4 - \partial_z^4 - 2\partial_t^2\partial_y^2 + 2\partial_x^2\partial_z^2 - 4\partial_t^2\partial_z\partial_y + 4\partial_t\partial_y^2\partial_z + 4\partial_t\partial_y\partial_z^2 - 4\partial_x\partial_y^2\partial_z.$$

*Example 30.* Following Examples 8 and 19 if $\mathscr{A} = \mathscr{H} \times \mathscr{H}$, then

$$\Delta_{\mathscr{A}} = \det \begin{bmatrix} \partial_t & \partial_x & 0 & 0 \\ \partial_x & \partial_t & 0 & 0 \\ 0 & 0 & \partial_y & \partial_z \\ 0 & 0 & \partial_z & \partial_y \end{bmatrix} = (\partial_t^2 - \partial_x^2)(\partial_y^2 - \partial_z^2).$$

*Example 31.* Following Examples 9 and 20 if $\mathscr{A} = \mathbb{C} \times \mathbb{C}$, then

$$\Delta_{\mathscr{A}} = \det \begin{bmatrix} \partial_t & -\partial_x & 0 & 0 \\ \partial_x & \partial_t & 0 & 0 \\ 0 & 0 & \partial_y & -\partial_z \\ 0 & 0 & \partial_z & \partial_y \end{bmatrix} = (\partial_t^2 + \partial_x^2)(\partial_y^2 + \partial_z^2).$$

*Example 32.* Following Examples 10 and 21 if $\mathscr{A} = \mathbb{H}$, then

$$\Delta_{\mathscr{A}} = \det \begin{bmatrix} \partial_t & -\partial_x & -\partial_y & -\partial_z \\ \partial_x & \partial_t & -\partial_z & \partial_y \\ \partial_y & \partial_z & \partial_t & -\partial_x \\ \partial_z & -\partial_y & \partial_x & \partial_t \end{bmatrix}.$$

We calculate:

$$\triangle_{\mathscr{A}} = \partial_t^4 + \partial_x^4 + \partial_y^4 + \partial_z^4 + 2\partial_t^2\partial_x^2 + 2\partial_t^2\partial_y^2 + 2\partial_t^2\partial_z^2 + 2\partial_x^2\partial_z^2 + 2\partial_x^2\partial_y^2 + 2\partial_y^2\partial_z^2.$$

*Example 33.* Following Examples 11 and 22 if $\mathscr{A} = \mathbb{R}_2$, the $2 \times 2$ real matrix algebra represented by $\mathbb{R}^4$, then

$$\triangle_{\mathscr{A}} = \det \begin{bmatrix} \partial_t & 0 & \partial_x & 0 \\ 0 & \partial_t & 0 & \partial_x \\ \partial_y & 0 & \partial_z & 0 \\ 0 & \partial_y & 0 & \partial_z \end{bmatrix} = \partial_t^2\partial_z^2 - \partial_x^2\partial_y^2.$$

### 8.4.2  Wagner's Laplace Equations vs. the $\mathscr{A}$-Laplace Equation

Let us briefly summarize the results of Wagner in [14]. The Laplace equation of complex variables is derived from the Cauchy Riemann equations $u_x = v_y$ and $u_y = -v_x$ as follows:

$$u_{xx} = (v_y)_x = (v_x)_y = (-u_y)_y \implies u_{xx} + u_{yy} = 0.$$

In short, if $f = u + iv$ is $\mathbb{C}$-differentiable, then both $u$ and $v$ must satisfy Laplace's equation. Wagner generalizes this calculation to a Frobenius algebra over $\mathbb{R}$. In particular, he derives from the symmetry of the partial derivatives that the Hessian matrix should be a paratrophic matrix (see the thesis by W. E. Deskins for definition and properties of such matrices [18]). He then argues that the Laplace equations can be read from the multiplication tables of Frobenius algebras. In Wagner's approach there is a set of $\frac{n(n-1)}{2}$ Laplace equations. When $n = 2$ we find precise agreement with Wagner for the semisimple examples. One of the great advantages to Wagner's equations is that a solution to Wagner's Laplace equations can be extended to an $\mathscr{A}$-differentiable function, although Wagner terms them *analytic* and we should emphasize his construction is given for the commutative case alone. In contrast, we do not attempt to develop conditions which allow us to extend a particular real-valued solution of the $\mathscr{A}$-Laplacian to a full $\mathscr{A}$-differentiable function. That is an interesting question, but we set it aside for future work.

For the interested reader, we exhibit this in a representative case to illustrate how our equation relates to Wagner's Laplace equations. The multiplication table for $\mathscr{A} = \mathbb{R} \oplus j\mathbb{R} \oplus j^2\mathbb{R}$ with $e_1 = 1, e_2 = j$ and $e_3 = j^2$ with $j^3 = 1$ is

| $\phantom{x}$ | 1 | $j$ | $j^2$ |
|---|---|---|---|
| 1 | 1 | $j$ | $j^2$ |
| $j$ | $j$ | $j^2$ | 1 |
| $j^2$ | $j^2$ | 1 | $j$ |

Wagner essentially argues that forcing the same pattern on the Hessian matrix $[\partial_{ij} u]$ yields the Laplace equations for the algebra. In this case Wagner's Laplace equations are:

$$(\partial_{xy} - \partial_{zz})u = 0, \qquad (\partial_{xz} - \partial_{yy})u = 0, \qquad (\partial_{yz} - \partial_{xx})u = 0.$$

In contrast, our $\mathscr{A}$-Laplacian from Example 27 can be factored and written as:

$$\triangle_{\mathscr{A}} = (\partial_x + \partial_y + \partial_z)(\partial_{xx} + \partial_{yy} + \partial_{zz} - \partial_{xy} - \partial_{yz} - \partial_{zx}).$$

Observe that the quadratic operator is formed by summing Wagner's Laplace operators. This is typical of the examples we have calculated. If we factor the $\mathscr{A}$-Laplacian, then the quadratic terms will correspond to sums of Wagner's Laplacians.

## 8.5 Conclusions and Future Work

The problem of generalizing complex variables has attracted the interest of mathematicians for over a century. While we have rediscovered some of these results, we make no claim to the originality of this work except in one regard. We believe the $\mathscr{A}$-Laplacian to be a new construction which we have yet to find in the vast literature on this subject. Certainly Wagner's construction of a set of Laplace equations is in many ways superior to our work, but his approach is limited to Frobenius algebras. In contrast, we have found the $\mathscr{A}$-Laplacian is solved by $\mathscr{A}$-differentiable functions even in the noncommutative case.

We intend to seek a general proof of Conjecture 1 in our next work. We already have some encouraging results from the higher dimensional real and complex matrix algebras. We also should mention that the problem of coordinate change warrants further attention; we have a preliminary proof of naturality of the $\mathscr{A}$-Laplacian in the commutative case. We find that if $\Phi : \mathscr{A} \to \mathscr{B}$ is an isomorphism, then $\triangle_{\mathscr{A}} = \det(\Phi)^2 \triangle_{\mathscr{B}}$. One goal of our next paper is to show the $\mathscr{A}$-Laplacian is natural with respect to algebra isomorphism for any semisimple algebra. The generalized Laplace equation applies to 9 of the 11 examples given in this paper. We do not know how to treat examples which are not semisimple. Certainly such cases are of interest to the literature. In fact, all of supermathematics concerns calculus over various generalizations of the dual numbers and this was the primary focus of the paper by Vladimirov and Volovich [13], which initially sparked our interest on generalized Cauchy Riemann equations. Finally, it would be interesting to obtain a harmonic function theory based on the $\mathscr{A}$-Laplacian much as is already known for Wagner's Laplace equations.

**Acknowledgments** We would like to express our gratitude to the organizers of The 8th Annual UNCG Regional Mathematics and Statistics Conference for providing a forum to express our research. We also thank Liberty University for a hospitable workspace during the Fall 2012 semester, the reviewer of this paper for several pedagogical improvements, and William J. Cook for useful insights into the algebra of our problem. We are of course responsible for any mistakes or oversights [1–18].

# References

1. Akivis, M.A., Rosenfeld, B.A.: Translations of mathematical monographs. Am. Math. Soc. **123** (1993)
2. Cartan, E.: Les groupes bilineaires et les systemes se nombres complexes, Oeuvres completes, partie 2. CNRS Paris, 7–105 (1984)
3. Dummit, D.S., Foote, R.M. (eds): Abstract Algebra. Wiley, Hoboken (2004)
4. Etingof, P., Golbert, O., Hensel, S., Liu, T., Schwendner, A., Vaintrob, D., Yudovina, E., Gerovitch, S.: Introduction to representation theory. Am. Math. Soc. **59** (2011)
5. Ferdinand Georg Frobenius: ber lineare Substitutionen und bilineare Formen. J. fr die reine und Angewandte Mathematik **84**, 1–63 (1878)
6. Hausdorff, F.: Zur Theorie der Systeme complexer Zahlen. Berichte ber die Verhandlugen der SŁchisischen Akademie der Wissenschaften zu Leipzig. Mathematisch-physikalische Klasse **52**, 43–61 (1900)
7. Ketchum, P.W.: Analytic functions of hypercomplex variables. Trans. Am. Math. Soc. **30**, 641–667 (1928)
8. Ludkovsky, S.V.: Differentiable functions of Cayley-Dickson numbers and line integration. Springer J. Math. Sci. **141**(3), 1231–1298 (2007)
9. Ringleb, F.: BeitrŁge zur Funktionentheorie in hyperkomplexen Systemen I. Rendiconti del Circolo Matematico di Palermo **57**, 311–340 (1933)
10. Rosenfeld, B.: Differentiable functions in associative and alternative algebras and smooth surfaces in projective spaces over these algebras. Publications De L'institut Mathematique. Nouvelle srie **62**(82), 67–71 (2000)
11. Scheffers, G.: Verallgemeinerung der Grundlagen der gewhnlich complexen Funktionen, I, II. Berichte ber die Verhandlugen der SŁchisischen Akademie der Wissenschaften zu Leipzig. Mathematisch-physikalische Klasse **46**, 120–134 (1894)
12. Spampinato, N.: Sulla Rappresentazione delle funzioni di variable bicomplessa totalmente derivabili. Annali di Matematica pura ed applicata **14**(4), 305–325 (1936)
13. Vladimirov, V.S., Volovich, I.V.: Superanalysis. V. A. Steklov Mathematics Institute. USSR Acad. Sci. **59**(1), 3–27 (1984)
14. Wagner, R.D.: The generalized Laplace equations in a function theory for commutative algebras. Duke Math. J. **15**, 455–461 (1948)
15. Ward, J.A.: A theory of analytic functions in linear associative algebras. Duke Math. J. **7**, 233–248 (1940)
16. Ward, J.A.: A theory of analytic functions in linear associative algebras. Doctoral Dissertation. University of Wisconsin (1939)
17. Ward, J.A.: From generalized Cauchy-Riemann equations to linear algebras. J. Am. Math. Soc. **4**(3), 456–461 (1953)
18. Deskins, W.E.: The Role of the Parastrophic Matrices in the Theory of Linear Associative Algebras. University of Wisconsin, Madison (1953)

# Chapter 9
# Fibonacci and Lucas Identities via Graphs

Joe DeMaio and John Jacobson

## 9.1 Introduction

Given a graph $G = (V, E)$, a set $S \subseteq V$ is an independent set of vertices if no two vertices in $S$ are adjacent. In our illustrations, we indicate membership in an independent set $S$ by shading the vertices in $S$. Let the set of all independent sets of a graph $G$ be denoted by $I(G)$ and let $i(G) = |I(G)|$. Note that $\emptyset \in I(G)$. The *path graph*, shown in Fig. 9.1, consists of the vertex set $V = \{v_1, v_2, \ldots, v_n\}$ and the edge set $E = \{\{v_1, v_2\}, \{v_2, v_3\}, \ldots, \{v_{n-1}, v_n\}\}$. The Fibonacci sequence is defined recursively as $F_n = F_{n-1} + F_{n-2}$ for positive integers $n \geq 2$ where $F_0 = 0$ and $F_1 = 1$ [6]. Table 9.1 shows the first few Fibonacci numbers.

In 1982, Prodinger and Tichy defined the Fibonacci number of a graph $G$, $i(G)$, to be the number of independent sets (including the empty set) of the graph $G$ [9]. They do so because the Fibonacci number of the path graph $P_n$ is the Fibonacci number $F_{n+2}$. The first few values of $i(P_n)$ are illustrated in Fig. 9.2.

In order to prove that $i(P_n) = F_{n+2}$ for all $n$, it must be shown that the number of independent sets on the path graph can be represented as a Fibonacci recurrence, that is $i(P_n) = i(P_{n-1}) + i(P_{n-2})$. First, partition $I(P_n)$ into two subsets: the set of all independent sets of vertices where vertex $n$ is not shaded and the set of all independent sets of vertices where the vertex $n$ is shaded. There are $i(P_{n-1})$ sequences of vertices that end with an unshaded vertex because an unshaded vertex can be added to any independent collection of vertices on the path graph of length $n - 1$ and still yield an independent set of vertices. Likewise there are $i(P_{n-2})$ sequences of vertices that end with a shaded vertex because two vertices, one unshaded and one shaded, can be added to every path graph with $n - 2$ vertices. Therefore $i(P_n) = i(P_{n-1}) + i(P_{n-2})$ and $i(P_n) = F_{n+2}$.

J. DeMaio (✉) • J. Jacobson
Department of Mathematics and Statistics,
Kennesaw State University, Kennesaw, GA 30144, USA
e-mail: jdemaio@kennesaw.edu; jjacob26@students.kennesaw.edu

J. Rychtář et al. (eds.), *Topics from the 8th Annual UNCG Regional Mathematics and Statistics Conference*, Springer Proceedings in Mathematics & Statistics 64, DOI 10.1007/978-1-4614-9332-7_9, © Springer Science+Business Media New York 2013

**Fig. 9.1** The path graph, $P_n$

**Table 9.1** Initial values of the Fibonacci sequence

| $n$ | 0 | 1 | 2 | 3 | 4 | 5 | 6 | 7 | 8 | 9 | 10 |
|-----|---|---|---|---|---|---|---|---|----|----|----|
| $F_n$ | 0 | 1 | 1 | 2 | 3 | 5 | 8 | 13 | 21 | 34 | 55 |

**Fig. 9.2** Independent sets on $P_1$, $P_2$, $P_3$, and $P_4$

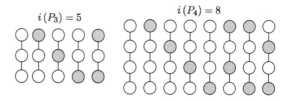

**Fig. 9.3** Independent sets on $C_3$ and $C_4$

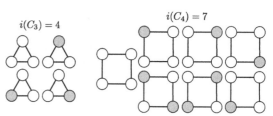

**Table 9.2** Initial values of the Lucas sequence

| $n$ | 0 | 1 | 2 | 3 | 4 | 5 | 6 | 7 | 8 | 9 | 10 |
|-----|---|---|---|---|---|----|----|----|----|----|-----|
| $L_n$ | 2 | 1 | 3 | 4 | 7 | 11 | 18 | 29 | 47 | 76 | 123 |

In [9], Prodinger and Tichy also determined the Fibonacci number of the cycle graph, $C_n$. Similar to the path graph, the values of $i(C_n)$ illustrated in Fig. 9.3 correlate to a recursive integer sequence, albeit one less well known than the Fibonacci sequence. The Lucas sequence is defined recursively as $L_n = L_{n-1} + L_{n-2}$ for positive integers $n \geq 2$ where $L_0 = 0$ and $L_1 = 1$ [6]. Table 9.2 shows the first few Lucas numbers. Accordingly, Prodinger and Tichy showed that the Fibonacci number of the cycle graph $C_n$ is the Lucas number $L_n$.

Since the publication of Prodinger and Tichy's 1982 paper, mathematicians have calculated the Fibonacci number of various graphs such as trees [5], an $M \times N$ lattice [3], and grids [2]. However, the relationship between independent sets and the Fibonacci sequence has not been used to combinatorially prove Fibonacci and Lucas identities. In *Proofs that Really Count*, Benjamin and Quinn offer combinatorial proofs for numerous Fibonacci and Lucas identities [1]. Nelson's *Proof Without Words* series [7, 8] provide purely visual arguments for several different types mathematical identities, some of which include the Fibonacci sequence. Here we join the concept of a visual proof using a graph with combinatorial methods to discover new identities for Fibonacci and Lucas numbers.

In order to realize the Fibonacci sequence, Benjamin and Quinn count the number of ways one can tile a $1 \times n$ board using square tiles with dimensions $1 \times 1$ and domino tiles with dimensions $1 \times 2$. We are able to prove many of the same identities using similar strategies on the path graph. However, there is a fundamental difference between tiling a board and constructing independent sets. Focusing on that difference, we are able to discover new identities.

Consider any two tilings. While we can append one tiling to the other, creating a larger one, we cannot break the tiling wherever we choose. Benjamin and Quinn restrict the breaking sites to the end of a square or the end of a domino. One cannot break the tiling in the middle of a domino. Now consider any path graph where an independent set of vertices is shaded. We can delete any edge to create two smaller path graphs each with an independent set of vertices. Although we can break a path graph wherever we please, we cannot join every pair of path graphs such that the resulting graph's shaded vertices form an independent set. For the remainder of this paper we will call two paths and their respective independent sets that can be joined to form an independent set on a larger path graph a *couple*.

## 9.2 Combinatorial Proofs of Fibonacci Identities by Means of the Path Graph

**Theorem 1.** *For $n \geq 3$, $F_{2n} = 2F_{n-1}F_n + F_n^2$.*

*Proof.* We know that there are $F_{2n}$ independent sets on the path graph $P_{2n-2}$. Now we partition $I(P_{2n-2})$ into three disjoint sets. Let $A$ be all independent sets on $P_{2n-2}$ that do not contain vertices $n-1$ and $n$, let $B$ be those that contain vertex $n$, and let $C$ be those that contain vertex $n-1$. There are $F_n^2$ independent sets in $A$ because while we exclude $n-1$ and $n$ from our count, we include all independent sets on the path including vertices 1 to $n-2$ and the path including vertices $n+1$ to $2n-2$ and $i(P_{n-2}) = F_n$. For set $B$, we must count the independent sets on the path from vertex 1 to $n-2$ and the path from vertex $n+2$ to $2n-2$. This gives us $F_n F_{n-1}$ independent sets in $B$. Similarly, set $C$ has $F_n F_{n-1}$ independent sets. Therefore, because $A \cup B \cup C = I(P_{2n-2})$, we have $F_{2n} = 2F_{n-1}F_n + F_n^2$. ∎

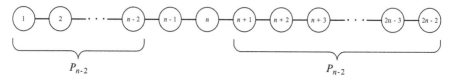

Independent sets in $A$

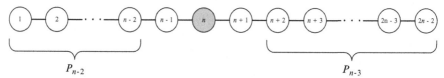

Independent sets in $B$

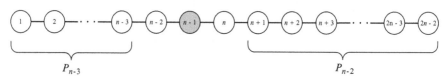

Independent sets in $C$

**Theorem 2.** *For* $n \geq 4$, $F_{3n+2} = F_{n+2}^3 - 2F_n^2 F_{n+2} + F_n^2 F_{n-2}$.

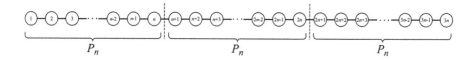

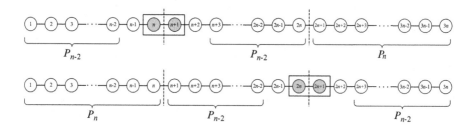

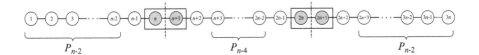

*Proof.* Consider the path graph $P_{3n}$. On the one hand, we know that $i(P_{3n}) = F_{3n+2}$. On the other hand, if we place three paths of length $n$ end to end, and count the number of independent sets on each path, we get $F_{n+2}^3$ different sets of shaded vertices in $P_{3n}$.

However, since $P_n$ and $P_n$ do not always form a couple, there are sets in $I(P_{3n})$ that, by definition, do not belong and must be removed. All of the sets that contain vertices $n$ and $n+1$ are not independent sets on $P_{3n}$, and neither are the sets containing vertices $2n$ and $2n+1$. So, we must remove $2F_n^2 F_{n+2}$ sets.

Through this subtraction, we have removed the sets containing vertices $n, n+1, 2n$, and $2n+1$ twice.

Therefore, by the inclusion–exclusion principle, we add $F_n^2 F_{n-2}$ sets which gives us the result

$$i(P_{3n}) = F_{n+2}^3 - 2F_n^2 F_{n+2} + F_n^2 F_{n-2}$$
$$= F_{3n+2}.$$

∎

Using the same proof technique, we are able to discover identities for $F_{4n+2}$ and $F_{5n+2}$ for $n \geq 4$.

$$F_{3n+2} = F_{n+2}^3 - 2F_n^2 F_{n+2} + F_n^2 F_{n-2} \tag{9.1}$$

$$F_{4n+2} = F_{n+2}^4 - 3F_n^2 F_{n+2}^2 + 2F_n^2 F_{n-2} F_{n+2} + F_n^4 - F_n^2 F_{n-2}^2 \tag{9.2}$$

$$F_{5n+2} = F_{n+2}^5 - 4F_n^2 F_{n+2}^3 + 3F_n^2 F_{n-2} F_{n+2}^2 + 3F_n^4 F_{n+2} - \tag{9.3}$$
$$2F_n^2 F_{n-2}^2 F_{n+2} - 2F_n^4 F_{n-2} + F_n^2 F_{n-2}^3$$

## 9.3 Combinatorial Proofs of Fibonacci and Lucas Identities by means of the Cycle Graph

In this section we prove a new identity relating Fibonacci and Lucas numbers. Recall that $i(C_n) = L_n$. As in the previous section, we are able to view this problem in terms of graphs and use the concept of a couple in our proofs.

**Theorem 3.** *For* $n \geq 4$, $L_{3n} = F_{n+2}^3 - 3F_n^2 F_{n+2} + 3F_n^2 F_{n-2} - F_{n-2}^3$.

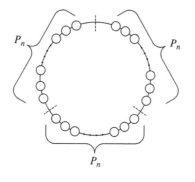

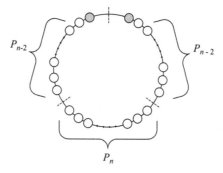

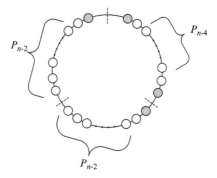

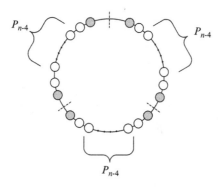

*Proof.* We know that there are $L_{3n}$ independent sets on $C_{3n}$. On the other hand, if we place three paths of length $n$ end to end, and count the number of independent sets on each path, we get $F_{n+2}^3$ different sets of shaded vertices in $C_{3n}$.

Although we have accounted for all independent sets on each individual path, this does not guarantee that each will be an independent set on $C_{3n}$. There are three instances where two paths do not form a couple.

So we subtract $3F_n^2 F_{n+2}$ from $F_{n+2}^2$. Now we must add back the number sets which we have removed twice.

Since this situation occurs three times, we add back $3F_n^2 F_{n-2}$ sets. Finally, by the inclusion–exclusion principle we must subtract $F_{n-2}^3$ sets.

Therefore, $L_{3n} = F_{n+2}^3 - 3F_n^2 F_{n+2} + 3F_n^2 F_{n-2} - F_{n-2}^3$. ∎

Using this method, we are able to discover new identities for $L_{4n}$ and $L_{5n}$ for $n \geq 4$.

$$L_{3n} = F_{n+2}^3 - 3F_n^2 F_{n+2} + 3F_n^2 F_{n-2} - F_{n-2}^3 \tag{9.4}$$

$$L_{4n} = F_{n+2}^4 - 4F_n^2 F_{n+2}^2 + 4F_n^2 F_{n-2} F_{n+2} + 2F_n^4 - 4F_n^2 F_{n-2}^2 + F_{n-2}^4 \tag{9.5}$$

$$L_{5n} = F_{n+2}^5 - 5F_n^2 F_{n+2}^3 + 5F_{n-2} F_n^2 F_{n+2}^2 + 5F_n^4 F_{n+2} - 5F_{n-2}^2 F_n^2 F_{n+2}$$

$$- 5F_{n-2} F_n^4 + 5F_{n-2}^3 F_n^2 - F_{n-2}^5 \tag{9.6}$$

## 9.4 Future Work

A minor shortcoming of using independent sets in graphs to represent the Fibonacci and Lucas sequences is the loss of the first few values of $n$ in these identities. We can realize $F_2$ by using the semi-controversial empty graph [4]. It is easy to state that only the empty set of vertices can be selected from the graph with no vertices and thus, $i(P_0) = F_2$. Finding a combinatorial realization for why $i(P_{-1}) = 1$ and $i(P_{-2}) = 0$ is less obvious. It is trivial to plug in specific values and show that the identities hold for these small $n$. However, it would be far more satisfying if a combinatorial interpretation for these non-existent graphs could be found.

The Fibonacci sequence and the Lucas sequence are famous examples of a more general integer sequence called the Gibonacci sequence [6]. For integers $G_0 = a$ and $G_1 = b$, the Gibonacci sequence is defined recursively as $G_n = G_{n-1} + G_{n-2}$ for positive integers $n \geq 2$. Because we are able to find graphs whose number of independent sets exhibit Fibonacci and Lucas recurrences, a natural next step in this work is to find graphs with a Gibonacci recurrence.

Generalizing the techniques explored in this paper in order to determine similar formulae for $F_{kn+2}$ and $L_{kn}$ represents an alternate and decidedly more challenging next step. In the identities presented above, the coefficients of the Fibonacci numbers are the number of different conflicts for path couples, and as $k$ grows in both $F_{kn+2}$ and $L_{kn}$, enumerating the possible combinations of adjacent and nonadjacent paths in a partitioned graph becomes very complex.

# References

1. Benjamin, A.T., Quinn, J.J.: Proofs That Really Count: The Art of Combinatorial Proof. The Mathematical Association of America, Washington, DC (2003)
2. Calkin, N.J., Wilf, H.S.: The number of independent sets in a grid graph. SIAM J. Discrete Math. **11**(1), 54–60 (1998, electronic)
3. Engel, K.: On the Fibonacci number of an $m \times n$ lattice. Fibonacci Quart. **28**(1), 72–78 (1990)
4. Harary, F., Read, R.: Is the null graph a pointless concept? In: Graphs and Combinatorics Conference, George Washington University. Springer, New York (1973)
5. Knopfmacher, A., Tichy, R.F., Wagner, S., Ziegler, V.: Graphs, partitions, and Fibonacci numbers. Discrete Appl. Math. **155**, 1175–1187 (2007)
6. Koshy, T.: Fibonacci and Lucas numbers with applications. Wiley, New York (2001)
7. Nelsen, R.B.: Proofs Without Words: Exercises in Visual Thinking. The Mathematical Association of America, Washington, DC (1993)
8. Nelsen, R.B.: Proofs Without Words. II: More Exercises in Visual Thinking. The Mathematical Association of America, Washington, DC (2001)
9. Prodinger, H., Tichy, R.: Fibonacci numbers of graphs. Fibonacci Quart. **20**(1),16–21 (1982)

# Chapter 10
# More Zeros of the Derivatives of the Riemann Zeta Function on the Left Half Plane

Ricky Farr and Sebastian Pauli

## 10.1 Introduction

Let $s \in \mathbb{C}$. We denote the real part of $s$ by $\sigma$ and the imaginary part of $s$ by $t$. For $\sigma > 1$ the Riemann zeta function $\zeta$ can be written as

$$\zeta(s) = \sum_{n=1}^{\infty} \frac{1}{n^s}. \tag{10.1}$$

By analytic continuation, $\zeta$ may be extended to the whole complex plane, with the exception of the simple pole $s = 1$. This analytic continuation is characterized by the functional equation

$$\zeta(1 - s) = 2\Gamma(s)\zeta(s)(2\pi)^{-s} \cos \frac{\pi s}{2}. \tag{10.2}$$

It follows directly from the functional equation (10.2) that $\zeta(-2j) = 0$ for all $j \in \mathbb{N}$. These zeros are called the real or trivial zeros of $\zeta$. Also, by the Prime Number Theorem, all nontrivial zeros must lie in the critical strip $0 \leq \sigma \leq 1$. By the Riemann hypothesis, the remaining (nontrivial) zeros of $\zeta$ are of the form $\frac{1}{2} + it$.

In this paper we numerically investigate the distribution of zeros of the derivatives $\zeta^{(k)}$ of $\zeta$ on the left half plane. The results of our computations, that considerably expand the list of previously published zeros [11, 15], can be found in Tables 10.1 and 10.2. For the rectangular region $-10 < \sigma < \frac{1}{2}$ and $|t| < 10$,

R. Farr (✉) • S. Pauli
Department of Mathematics and Statistics, University of North Carolina Greensboro, Greensboro, NC 27402, USA
e-mail: refarr@uncg.edu; s_pauli@uncg.edu

J. Rychtář et al. (eds.), *Topics from the 8th Annual UNCG Regional Mathematics and Statistics Conference*, Springer Proceedings in Mathematics & Statistics 64, DOI 10.1007/978-1-4614-9332-7_10, © Springer Science+Business Media New York 2013

**Table 10.1** The number of zeros of $\zeta^{(k)}(\sigma + it)$ with $k \leq 32$ in $-10 < \sigma < 0$, $|t| < 10$, the number of complex conjugate pairs of non-real zeros, and the number of real zeros in this region

| | # of zeros of $\zeta^{(k)}(\sigma + it)$ | | | Zeros of $\zeta^{(k)}(\sigma + it)$ | | | | $0 < \sigma < 1/2$ |
| | $-10 < \sigma < 0$ | | | $-10 < \sigma < 0$ | | | | |
| $k$ | $|t| < 10$ | $0 < t < 10$ | $t = 0$ | $t = 0$ | | | | $|t| < 10$ |
|---|---|---|---|---|---|---|---|---|
| 0 | 4 | 0 | 4 | $-2$ | $-4$ | $-6$ | $-8$ | |
| 1 | 3 | 0 | 3 | $-2.7173$ | $-4.9368$ | $-7.0746$ | | |
| 2 | 5 | 1 | 3 | $-3.5958$ | $-6.0290$ | $-8.2786$ | | |
| 3 | 5 | 2 | 3 | $-4.7157$ | $-7.2920$ | $-9.6047$ | | |
| 4 | 6 | 2 | 2 | $-6.1265$ | $-8.7016$ | | | |
| 5 | 5 | 2 | 1 | $-7.7119$ | | | | $0.2876 \pm 4.6944i$ |
| 6 | 7 | 2 | 3 | $-4.3284$ | $-6.6083$ | $-9.3445$ | | |
| 7 | 8 | 3 | 2 | $-5.6191$ | $-8.4425$ | | | |
| 8 | 7 | 3 | 1 | $-7.5186$ | | | | $0.4183 \pm 5.4753i$ |
| 9 | 9 | 3 | 3 | $-4.7059$ | $-6.5553$ | $-9.3794$ | | |
| 10 | 10 | 4 | 2 | $-5.7309$ | $-8.5500$ | | | |
| 11 | 9 | 4 | 1 | $-7.7120$ | | | | $0.4106 \pm 6.1502i$ |
| 12 | 11 | 4 | 3 | $-5.1849$ | $-6.8533$ | $-9.6751$ | | |
| 13 | 12 | 5 | 2 | $-6.1124$ | $-8.9100$ | | | |
| 14 | 11 | 5 | 1 | $-8.1400$ | | | | $0.3447 \pm 6.7636i$ |
| 15 | 12 | 5 | 2 | $-5.6697$ | $-7.3600$ | | | |
| 16 | 14 | 6 | 2 | $-6.6469$ | $-9.4393$ | | | |
| 17 | 13 | 6 | 1 | $-8.7229$ | | | | $0.2494 \pm 7.3344i$ |
| 18 | 14 | 6 | 2 | $-6.1556$ | $-8.0019$ | | | |
| 19 | 15 | 7 | 1 | $-7.3040$ | | | | |
| 20 | 15 | 7 | 1 | $-9.4151$ | | | | $0.1378 \pm 7.8732$ |
| 21 | 16 | 7 | 2 | $-6.6561$ | $-8.7394$ | | | |
| 22 | 17 | 8 | 1 | $-8.0675$ | | | | |
| 23 | 16 | 8 | 0 | | | | | $0.0163 \pm 8.3861i$ |
| 24 | 18 | 8 | 2 | $-7.1929$ | $-9.5491$ | | | $0.4681 \pm 8.7645i$ |
| 25 | 19 | 9 | 1 | $-8.9089$ | | | | |
| 26 | 20 | 9 | 2 | $-7.3618$ | $-8.2504$ | | | |
| 27 | 19 | 9 | 1 | $-7.8131$ | | | | $0.3116 \pm 9.244i$ |
| 28 | 21 | 10 | 1 | $-9.8049$ | | | | |
| 29 | 22 | 10 | 2 | $-7.7492$ | $-9.1919$ | | | |
| 30 | 21 | 10 | 1 | $-8.6103$ | | | | $0.1516 \pm 9.7083i$ |
| 31 | 22 | 11 | 0 | | | | | |
| 32 | 23 | 11 | 1 | $-8.2087$ | | | | |

Furthermore, the real zeros in this region and the zeros in the strip $0 < \sigma < \frac{1}{2}$, $|t| < 10$ are given to four decimal digits

**Table 10.2**  All zeros of $\zeta^{(k)}(\sigma + it)$ with $k \leq 29$ in $-10 < \sigma < 0, 0 < |t| < 10$

| $k$ | # | Zeros of $\zeta^{(k)}(\sigma + it)$ with $-10 < \sigma < 0$ and $0 < |t| < 10$ | | | |
|---|---|---|---|---|---|
| 2 | 1 | $-0.3551 \pm 3.5908i$ | | | |
| 3 | 1 | $-2.1101 \pm 2.5842i$ | | | |
| 4 | 2 | $-0.8375 \pm 3.8477i$ | $-3.2403 \pm 1.6896i$ | | |
| 5 | 2 | $-2.1841 \pm 3.0795i$ | $-4.2739 \pm 0.6624i$ | | |
| 6 | 2 | $-1.2726 \pm 4.0742i$ | $-3.1694 \pm 2.2894i$ | | |
| 7 | 3 | $-0.4133 \pm 4.8453i$ | $-2.3934 \pm 3.4063i$ | $-3.8750 \pm 1.4918i$ | |
| 8 | 3 | $-1.6703 \pm 4.2784i$ | $-3.2523 \pm 2.7170i$ | $-4.5682 \pm 0.8112i$ | |
| 9 | 3 | $-0.9672 \pm 4.9985i$ | $-2.6410 \pm 3.6749i$ | $-3.9459 \pm 2.0452i$ | |
| 10 | 4 | $-0.2748 \pm 5.6133i$ | $-2.0391 \pm 4.4684i$ | $-3.4229 \pm 3.0609i$ | $-4.5121 \pm 1.3321i$ |
| 11 | 4 | $-1.4413 \pm 5.1493i$ | $-2.9062 \pm 3.9132i$ | $-4.0769 \pm 2.4384i$ | $-5.0310 \pm 0.7641i$ |
| 12 | 4 | $-0.8452 \pm 5.7473i$ | $-2.3874 \pm 4.6486i$ | $-3.6307 \pm 3.3459i$ | $-4.6218 \pm 1.8307i$ |
| 13 | 5 | $-0.2500 \pm 6.2811i$ | $-1.8653 \pm 5.2971i$ | $-3.1788 \pm 4.1283i$ | $-4.2445 \pm 2.7740i$, |
| | | $-5.1019 \pm 1.1817i$ | | | |
| 14 | 5 | $-1.3402 \pm 5.8783i$ | $-2.7202 \pm 4.8199i$ | $-3.8543 \pm 3.5969i$ | $-4.7812 \pm 2.1996i$, |
| | | $-5.5404 \pm 0.6780i$ | | | |
| 15 | 5 | $-0.8124 \pm 6.4056i$ | $-2.2551 \pm 5.4415i$ | $-3.4521 \pm 4.3265i$ | $-4.4411 \pm 3.0614i$, |
| | | $-5.2367 \pm 1.6383i$ | | | |
| 16 | 6 | $-0.2827 \pm 6.8886i$ | $-1.7845 \pm 6.0069i$ | $-3.0400 \pm 4.9834i$ | $-4.0887 \pm 3.8241i$, |
| | | $-4.9528 \pm 2.5231i$ | $-5.6490 \pm 1.0311i$ | | |
| 17 | 6 | $-1.3092 \pm 6.5262i$ | $-2.6197 \pm 5.5821i$ | $-3.7242 \pm 4.5121i$ | $-4.6486 \pm 3.3161i$, |
| | | $-5.4130 \pm 1.9836i$ | $-6.0680 \pm 0.5743i$ | | |
| 18 | 6 | $-0.8299 \pm 7.0068i$ | $-2.1924 \pm 6.1331i$ | $-3.3491 \pm 5.1402i$ | $-4.3279 \pm 4.0324i$, |
| | | $-5.1468 \pm 2.8068i$ | $-5.8098 \pm 1.4611i$ | | |
| 19 | 7 | $-0.3475 \pm 7.4543i$ | $-1.7592 \pm 6.6440i$ | $-2.9648 \pm 5.7192i$ | $-3.9939 \pm 4.6871i$, |
| | | $-4.8654 \pm 3.5483i$ | $-5.5889 \pm 2.2963i$ | $-6.1583 \pm 0.8859i$ | |
| 20 | 7 | $-1.3211 \pm 7.1206i$ | $-2.5729 \pm 6.2569i$ | $-3.6489 \pm 5.2913i$ | $-4.5694 \pm 4.2268i$, |
| | | $-5.3472 \pm 3.0608i$ | $-5.9945 \pm 1.7820i$ | $-6.6140 \pm 0.4394i$ | |
| 21 | 7 | $-0.8787 \pm 7.5677i$ | $-2.1744 \pm 6.7594i$ | $-3.2944 \pm 5.8530i$ | $-4.2605 \pm 4.8536i$, |
| | | $-5.0870 \pm 3.7617i$ | $-5.7837 \pm 2.5734i$ | $-6.3545 \pm 1.2934i$ | |
| 22 | 8 | $-0.4328 \pm 7.9887i$ | $-1.7703 \pm 7.2313i$ | $-2.9319 \pm 6.3785i$ | $-3.9406 \pm 5.4371i$, |
| | | $-4.8118 \pm 4.4095i$ | $-5.5554 \pm 3.2943i$ | $-6.1750 \pm 2.0870i$ | $-6.6413 \pm 0.7581i$ |
| 23 | 8 | $-1.3613 \pm 7.6765i$ | $-2.5625 \pm 6.8727i$ | $-3.6113 \pm 5.9836i$ | $-4.5240 \pm 5.0128i$, |
| | | $-5.3115 \pm 3.9611i$ | $-5.9806 \pm 2.8250i$ | $-6.5366 \pm 1.5912i$ | $-7.1892 \pm 0.1700i$ |
| 24 | 8 | $-0.9481 \pm 8.0980i$ | $-2.1871 \pm 7.3395i$ | $-3.2737 \pm 6.4980i$ | $-4.2254 \pm 5.5784i$, |
| | | $-5.0539 \pm 4.5827i$ | $-5.7671 \pm 3.5097i$ | $-6.3712 \pm 2.3553i$ | $-6.8798 \pm 1.1259i$ |
| 25 | 9 | $-0.5313 \pm 8.4984i$ | $-1.8064 \pm 7.7820i$ | $-2.9291 \pm 6.9843i$ | $-3.9174 \pm 6.1112i$, |
| | | $-4.7841 \pm 5.1658i$ | $-5.5378 \pm 4.1485i$ | $-6.1844 \pm 3.0574i$ | $-6.7253 \pm 1.8906i$, |
| | | $-7.1206 \pm 0.6504i$ | | | |
| 26 | 9 | $-0.1113 \pm 8.8798i$ | $-1.4211 \pm 8.2028i$ | $-2.5782 \pm 7.4458i$ | $-3.6013 \pm 6.6153i$, |
| | | $-4.5038 \pm 5.7155i$ | $-5.2952 \pm 4.7478i$ | $-5.9817 \pm 3.7117i$ | $-6.5664 \pm 2.6042i$, |
| | | $-7.0463 \pm 1.4126i$ | | | |
| 27 | 9 | $-1.0318 \pm 8.6041i$ | $-2.2218 \pm 7.8850i$ | $-3.2780 \pm 7.0941i$ | $-4.2144 \pm 6.2361i$, |
| | | $-5.0410 \pm 5.3132i$ | $-5.7647 \pm 4.3261i$ | $-6.3901 \pm 3.2731i$ | $-6.9206 \pm 2.1489i$, |
| | | $-7.3814 \pm 0.9448i$ | | | |

(continued)

**Table 10.2**  (contnued)

| # |    |                      |                      |                      |                      |
|---|----|----------------------|----------------------|----------------------|----------------------|
| 28 | 10 | $-0.6389 \pm 8.9878i$ | $-1.8606 \pm 8.3044i$ | $-2.9484 \pm 7.5503i$ | $-3.9169 \pm 6.7308i,$ |
|    |    | $-4.7767 \pm 5.8489i$ | $-5.5353 \pm 4.9061i$ | $-6.1978 \pm 3.9018i$ | $-6.7680 \pm 2.8338i,$ |
|    |    | $-7.2490 \pm 1.7019i$ | $-7.6182 \pm 0.5486i$ |                      |                      |
| 29 | 10 | $-0.2428 \pm 9.3554i$ | $-1.4951 \pm 8.7056i$ | $-2.6132 \pm 7.9860i$ | $-3.6122 \pm 7.2024i,$ |
|    |    | $-4.5034 \pm 6.3583i$ | $-5.2947 \pm 5.4558i$ | $-5.9918 \pm 4.4954i$ | $-6.5986 \pm 3.4759i,$ |
|    |    | $-7.1165 \pm 2.3954i$ | $-7.5353 \pm 1.2495i$ |                      |                      |

The column # contains the number of conjugate pairs of zeros. All zeros listed are simple and are rounded to four decimal digits. It is expected that both the real and imaginary parts of the zeros are transcendental and linearly independent of each other

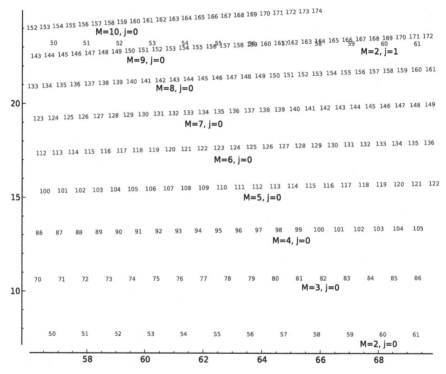

**Fig. 10.1** The zeros of $\zeta^{(k)}(\sigma + it)$ for $52 < \sigma < 70$, $0 < t < 25$, where k denotes a zero of $\zeta^{(k)}$. The conjectured chains of zeros are labeled by $M$ and $j$ (compare Theorem 3)

Table 10.1 contains the number of zeros of $\zeta^{(k)}$, its real zeros, and its zeros with $0 < \sigma < \frac{1}{2}$. Table 10.2 contains non-real zeros with $\sigma < 0$ in that region. We find that some of the conjectured chains of zeros of the derivatives on the right half plane [2, 9] (see Fig. 10.1) appear to continue to the left half plane which is illustrated in Fig. 10.3.

We first recall results about the distribution of the zeros of $\zeta^{(k)}$ on the right half plane (Sect. 10.2) and the left half plane (Sect. 10.3). Section 10.4 contains a description of the methods we used to evaluate $\zeta^{(k)}$. It is followed by a discussion of the methods that we used to find the zeros of $\zeta^{(k)}$ in Sect. 10.5.

## 10.2  Zeros on the Right Half Plane

Assuming the Riemann Hypothesis, the non-real zeros of $\zeta$ are all on the critical line $\sigma = \frac{1}{2}$, while the non-real zeros of $\zeta^{(k)}$ appear to be distributed mostly to the right of the critical line with some outliers located to its left.

### 10.2.1  Zeros with $0 < \sigma < \frac{1}{2}$

Speiser related the Riemann Hypothesis to the distribution of zeros of the first derivative.

**Theorem 1 (Speiser [10]).** *The Riemann Hypothesis is equivalent to $\zeta'(s)$ having no zeros in $0 < \sigma < \frac{1}{2}$.*

A simpler and more instructive proof of this result was given by Levinson and Montgomery [8]. They also proved, assuming the Riemann Hypothesis, that $\zeta^{(k)}(s)$ has at most a finite number of non-real zeros with $\sigma < \frac{1}{2}$, for fixed $k \geq 2$.

**Theorem 2 (Yıldırım [15]).** *The Riemann Hypothesis implies that $\zeta''$ and $\zeta'''$ have no zeros in the strip $0 \leq \sigma \leq \frac{1}{2}$.*

The Riemann Hypothesis also implies that $\zeta^{(k)}$ for $k > 0$ has only finitely many zeros in $0 \leq \sigma \leq \frac{1}{2}$ [8].

Our computations show that higher derivatives have zeros in this strip, see Table 10.1. Because of the very well-defined and predictable patterns in the distribution of the zeros of $\zeta^{(k)}$ in Fig. 10.2, we expect that the zeros listed in the table are the only zeros of $\zeta^{(k)}$ for $k \leq 32$.

### 10.2.2  Zeros with $\sigma > \frac{1}{2}$

The real parts of the zeros of $\zeta^{(k)}$ can be effectively bounded from above by absolute constants. For $\zeta'$ and $\zeta''$ Skorokhdov [9] gives the bounds:

$$\zeta'(\sigma + it) \neq 0 \quad \text{for} \quad \sigma > 2.93938,$$

$$\zeta''(\sigma + it) \neq 0 \quad \text{for} \quad \sigma > 4.02853.$$

For $k \geq 3$ such general upper bounds were given by Spira [11] and later improved by Verma and Kaur [14]:

$$\zeta^{(k)}(\sigma + it) \neq 0 \quad \text{for} \quad \sigma > q_2 k + 2,$$

where $q_2$ is given by the formula

$$q_M = \frac{\log\left(\frac{\log M}{\log(M+1)}\right)}{\log\left(\frac{M}{M+1}\right)}.$$

Spira [11] computed zeros of the first and second derivatives of $\zeta(s)$ for $0 < t < 100$ and noticed that they occur in pairs. Skorokhodov [9] went further in his computation and noticed that the zeros of derivatives of $\zeta$ seem to form chains, that is for each zero $z^{(k)}$ of $\zeta^{(k)}$ there seems to be a corresponding zero $z^{(k+1)}$ of $\zeta^{(k+1)}$. Indeed, for sufficiently large $k$ the existence of these chains is a direct consequence of the following theorem.

**Theorem 3 (Binder et al. [2]).** *Let $M \geq 2$ be an integer and let $u$ be a solution of $1 - \frac{1}{e^u - 1} - \frac{1}{e^u}\left(1 + \frac{1}{u}\right) \geq 0$, that is, $u \geq 1.1879\ldots$. If $k > \frac{u(2M+3)}{q_M - q_{M+1}}$, then for each $j \in \mathbb{Z}$ the rectangular region consisting of all $s = \sigma + it$ with*

$$q_M k - (M+1)u < \sigma < q_M k + (M+1)u \tag{10.3}$$

*and*

$$\frac{2\pi j}{\log(M+1) - \log(M)} < t < \frac{2\pi(j+1)}{\log(M+1) - \log(M)}, \tag{10.4}$$

*contains exactly one zero of $\zeta^{(k)}$. This zero is simple.*

So, given $M \geq 2$, $j \in \mathbb{Z}$ and $l > \frac{u(2M+3)}{q_M - q_{M+1}}$ for the zero of $\zeta^{(l)}$ in the region determined by (10.3) and (10.4) for $k = l$ there is a corresponding zero of $\zeta^{(l+1)}$ in the region determined by (10.3) and (10.4) for $k = l + 1$. Figure 10.1 illustrates the phenomenon of the chains of zeros of derivatives of $\zeta$. The zeros shown in the chains labeled $M = 2$, $j = 0$ and $M = 2$, $j = 1$ are in the rectangular regions from Theorem 3 and the zeros in the chain labeled $M = 3$, $j = 1$ are in the regions for $M = 3$ and $j = 1$ starting at the 77th derivative. The other chains are labeled by the parameters $M$ and $j$ of the regions into which higher derivatives in the chains eventually fall farther to the right.

## 10.3    Zeros on the Left Half Plane

It follows immediately from the functional equation (10.2) that $\zeta(s) = 0$ for $s = -2n$ where $n \in \mathbb{N}$. The zeros of the first derivative are the zeros postulated by the theorem of Rolle.

**Theorem 4 (Levinson and Montgomery [8]).** *For $n \geq 2$ there is exactly one zero of $\zeta'$ in the interval $(-2n, -2n + 2)$ and there are no other zeros of $\zeta'$ with $\sigma \leq 0$.*

Unlike on the right half plane, on the left there is no general left bound for the non-real zeros of $\zeta^{(k)}$. Spira showed:

**Theorem 5 (Spira [12]).** *For $k > 0$ there is an $\alpha_k$ so that $\zeta^{(k)}$ has only real zeros for $\sigma < \alpha_k$, and exactly one real zero in each open interval $(-2n - 1, -2n + 1)$ for $1 - 2n < \alpha_k$.*

The location of a non-real zero of the second derivative on the left half plane shows up in [11]. For both $\zeta''(s)$ and $\zeta'''(s)$ Yıldırım [15] proved the existence of exactly one pair of conjugate nontrivial zeros with $\sigma < 0$ and gave their location.

**Theorem 6 (Levinson and Montgomery [8]).** *If $\zeta^{(k)}$ has only a finite number of non-real zeros in $\sigma < 0$, then $\zeta^{(k+1)}$ has the same property.*

Hence, the absolute value of the non-real zeros of $\zeta^{(k)}$ on the left half plane can be bounded. This can be done by iteratively generalizing Yıldırım's methods for the second and third derivatives to higher derivatives.

Table 10.2 contains all the zeros of $\zeta^{(k)}(\sigma + it)$ with $-10 < \sigma < 0, 0 < |t| < 10$ for $2 \leq k \leq 29$. The patterns of the distribution of zeros in Fig. 10.2 suggest that these are all the zeros for these derivatives on the left half plane.

## 10.4    Evaluating $\zeta^{(k)}$ on the Left Half Plane

Methods for evaluating $\zeta$ and $\zeta^{(k)}$ include Euler–Maclaurin summation (see, for example, [4]) or convergence acceleration for alternating sums [3]. Implementations for the evaluation of $\zeta$ can be found in various computer algebra systems. The Python library mpmath [6] contains functions for evaluating derivatives of Hurwitz zeta functions, and thus $\zeta^{(k)}$, on the right half plane using Euler–Maclaurin summation.

We considered two different approaches for evaluating $\zeta^{(k)}$ in the left half plane. Because of speed and ease of implementation we use Euler–Maclaurin summation rather than the derivatives of the functional equation (see [1] for formulas for these). Using Euler–Maclaurin summation we obtain for $\sigma = \Re(s) > 1$ that

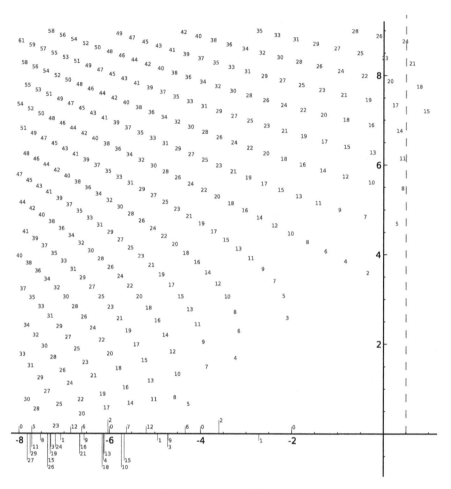

**Fig. 10.2** The zeros of $\zeta(\sigma + it)$ and its derivatives $\zeta^{(k)}(\sigma + it)$ for $k \leq 61$ in $-10 < \sigma < 1$, $0 < t < 9$, where 0 denotes a zero of $\zeta$ and k denotes a zero of $\zeta^{(k)}$. All zeros are simple

$$(-1)^k \zeta^{(k)}(s) = \sum_{n=2}^{\infty} \frac{\log^k(n)}{n^s}$$

$$= \sum_{n=2}^{N-1} \frac{\log^k(n)}{n^s} + \sum_{n=N}^{\infty} \frac{\log^k(n)}{n^s}$$

$$= \sum_{n=2}^{N-1} \frac{\log^k(n)}{n^s} + \int_{N}^{\infty} \frac{\log^k(x)}{x^s} dx + \frac{1}{2} \frac{\log^k(N)}{N^s}$$

$$+ \sum_{j=1}^{v} \frac{B_{2j}}{(2j)!} \frac{d^{2j-1}}{dx^{2j-1}} \frac{\log^k(x)}{x^s} \Big|_{x=N}^{\infty} + R_{2v}$$

$$= \sum_{n=2}^{N-1} \frac{\log^k(n)}{n^s} + \int_N^\infty \frac{\log^k(x)}{x^s}dx + \frac{1}{2}\frac{\log^k(N)}{N^s}$$

$$- \sum_{j=1}^{v} \frac{B_{2j}}{(2j)!}\frac{d^{2j-1}}{dx^{2j-1}}\frac{\log^k(x)}{x^s}\bigg|_{x=N} + R_{2v},$$

where $N \in \mathbb{N}^{>2}$, $v \in \mathbb{N}^{>2}$, and $R_{2v}$ is the error term. Repeated integration by parts yields:

$$\int_N^\infty \frac{\log^k(x)}{x^s}dx = \frac{\log^k(N)}{(s-1)N^{s-1}} \sum_{r=0}^{k} \frac{k!}{(k-r)!}\frac{\log^{-r}(N)}{(s-1)^r}.$$

Thus,

$$(-1)^k \zeta^{(k)}(s) = \sum_{n=2}^{N-1} \frac{\log^k(n)}{n^s} + \frac{\log^k(N)}{(s-1)N^{s-1}} \sum_{r=0}^{k} \frac{k!}{(k-r)!}\frac{\log^{-r}(N)}{(s-1)^r} + \frac{1}{2}\frac{\log^k(N)}{N^s}$$

$$- \sum_{j=1}^{v} \frac{B_{2j}}{(2j)!}\frac{d^{2j-1}}{dx^{2j-1}}\frac{\log^k(x)}{x^s}\bigg|_{x=N} + R_{2v},$$

$$(10.5)$$

The error term $R_{2v}$ is given by

$$R_{2v} = \frac{1}{(2v)!}\int_N^\infty \hat{B}_{2v}(x) f^{(2v)}(x)dx,$$

with $f(x) = \frac{\log^k(x)}{x^s}$ as discussed in [4]. We use the non-central Stirling numbers of the first kind (see [5]), to represent the derivatives of $f$. The non-central Stirling numbers of the first kind $S(r, i, s)$ satisfy the recurrence

$$S(1, 0, s) = -s, S(1, 1, s) = 1$$
$$S(r+1, 0, s) = (-s-r)S(r, 0, s)$$
$$S(r+1, i, s) = (-s-r)S(r, i, s) + S(r, i-1, s), 1 \le i \le r$$
$$S(r+1, r+1, s) = S(r, r, s).$$

With these the derivatives of $f$ can be written as

$$f^{(r)}(x) = x^{-s-r} \sum_{i=0}^{r} S(r, i, s)(k)_i \log^{k-i}(x),$$

where $(k)_i$ denotes the $i$-th falling factorial of $k$ [5].

We now bound the error term, $R_{2v}$. Observe that

$$|R_{2v}| = \left| \frac{1}{(2v)!} \int_N^\infty \hat{B}_{2v}(x) f^{(2v)}(x) dx \right| \tag{10.6}$$

$$\leq \frac{|B_{2v}|}{(2v)!} \int_N^\infty |f^{(2v)}(x)| dx \tag{10.7}$$

$$= \frac{|B_{2v}|}{(2v)!} \int_N^\infty \left| x^{-s-2v} \sum_{i=0}^{2v} S(2v, i, s)(k)_i \log^{k-i}(x) \right| dx \tag{10.8}$$

$$\leq \frac{|B_{2v}|}{(2v)!} \sum_{i=0}^{2v} \int_N^\infty \left| S(2v, i, s)(k)_i \frac{\log^{k-i}(x)}{x^{s+2v}} \right| dx \tag{10.9}$$

$$= \frac{|B_{2v}|}{(2v)!} \sum_{i=0}^{2v} |S(2v, i, s)|(k)_i \int_N^\infty \frac{\log^{k-i}(x)}{x^{\sigma+2v}} dx \tag{10.10}$$

$$\leq \frac{|B_{2v}|}{(2v)!} \sum_{i=0}^{2v} |S(2v, i, s)|(k)_i \left( \int_N^\infty \frac{\log^k(x)}{x^{\sigma+2v}} dx \right). \tag{10.11}$$

The error term $R_{2v}$ converges for $\sigma + 2v > 1$ and $N \in \mathbb{N}^{>2}$, thus (10.5) can be used to evaluate $\zeta^{(k)}$ for $\sigma > 1 - 2v$. Since we are evaluating $\zeta^{(k)}$ on a bounded region with $|\sigma| \leq 10$ the error can be bounded by (10.11) on the entire region. We set $v = 101$, which yields $\sigma + 2v > 1$ in the region and gives a good balance of the values for $v$ and $N$. To determine the value $N$ should take, we evaluate the bound given above for $N = 200, 300, \ldots$ until the error is as small as desired. For example, if $s = -10 + 10i, k = 100, v = 101$, and $N = 200$, then $|R_{2v}| < 1.769892 \cdot 10^{-100}$. If $N = 1500$, then $|R_{2v}| < 1.245704 \cdot 10^{-253}$.

## 10.5 Finding Zeros

We found the zeros on the left half plane by following the chains of zeros of derivatives of $\zeta$ from the right half plane (see Figs. 10.1 and 10.3). For given $M \geq 2$, $j \in \mathbb{Z}$, and sufficiently large $k$ the center

$$s = q_M k + \frac{2\pi(j + 0.5)}{\log(M + 1) - \log(M)}$$

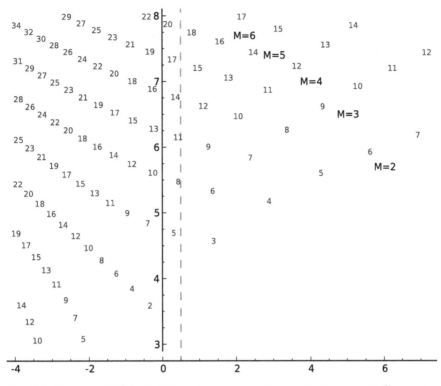

**Fig. 10.3** The zeros of $\zeta^{(k)}(\sigma + it)$ for $-4 < \sigma < 8$ and $3 < t < 8$. The zeros of $\zeta^{(k)}$ are at the center of the numbers k. The first five chains of zeros that we followed from the right to the left half plane are labeled $M = 2, \ldots, M = 6$ (see Sect. 10.2)

of the rectangular region from Theorem 3 is a good approximation to the zero in this region which we improved using Newton's method.

Now assume that we know a zero $z_M^{(k)}$ of $\zeta^{(k)}$ and a zero $z_M^{(k+1)}$ of $\zeta^{(k+1)}$ in the chain given by some $M$ and $j$. We used

$$s = z_M^{(k)} - \left( z_M^{(k+1)} - z_M^{(k)} \right)$$

as a first approximation for the zero of $\zeta^{(k-1)}$ in that chain, which again was improved with Newton's method.

We assured that we had found all zeros of $\zeta^{(k)}$ with $1 \leq k \leq 61$ in $-10 < \sigma < \frac{1}{2}$, $|t| < 10$ by counting the zeros using contour integration. The only pole of $\zeta^{(k)}$ is at one and thus outside our region of interest. So for any simple closed contour $C$ in $-10 < \sigma < \frac{1}{2}$, $|t| < 10$, by the argument principle, the number of zeros of $\zeta^{(k)}$ inside $C$ counted as many times as their multiplicity is

$$n = \frac{1}{2\pi i} \int_C \left( \frac{\zeta^{(k+1)}}{\zeta^{(k)}} \right)(s)\, ds.$$

For $1 \le k \le 61$ we counted the zeros of $\zeta^{(k)}$ by integrating along the border of the rectangular region $-10 < \sigma < \frac{1}{2}$, $|t| < 10$. We also integrated along the sides of a square region with side length $10^{-6}$ centered around each approximation $z$ of the zeros to make sure that this region contained exactly one simple zero.

All computations and plotting were conducted with the computer algebra system Sage [13]. We evaluated $\zeta^{(k)}$ with our implementation of the method described in Sect. 10.4 which was verified, on the right half plane, with the Hurwitz zeta function in mpmath [6] and our implementation of $\zeta^{(k)}$ based on convergence acceleration for alternating series. For the integration we used the numerical integration function of Sage which calls the GNU Scientific Library [7] using an adaptive Gauss–Kronrod rule.

# References

1. Apostol, T.M.: Formulas for higher derivatives of the Riemann zeta function. Math. Comp. **44**(169), 223–232 (1985)
2. Binder, T., Pauli, S., Saidak, F.: New zero free regions for the derivatives of the Riemann Zeta function, Rocky Mountain Journal of Mathematics (2011)
3. Cohen, H., Villegas, F.R., Zagier, D.: Convergence acceleration of alternating series. Exp. Math. **9**, 3–12 (2000)
4. Edwards, H.M.: Riemann's Zeta function. In: Pure and Applied Mathematics, vol. 58. Academic, New York (1974)
5. Janjic, M.: On non-central stirling numbers of the first kind. http://adsabs.harvard.edu/abs/2009arXiv0901.2655J (2009)
6. Joansson, F., et al.: Mpmath: a Python library for arbitrary-precision floating-point arithmetic. http://code.google.com/p/mpmath/ (2010)
7. Jungman, G., Gough, B., et al.: GSL - GNU Scientific Library. http://www.gnu.org/software/gsl/ (2011)
8. Levinson, N., Montgomery, H.L.: Zeros of the derivatives of the Riemann Zeta function. Acta Math. **133**, 49–65 (1974)
9. Skorokhodov, S.L.: Padé approximants and numerical analysis of the Riemann Zeta function. Zh. Vychisl. Mat. Mat. Fiz. **43**(9), 1330–1352 (2003)
10. Speiser, A.: Geometrisches zur Riemannschen Zetafunktion. Math. Ann. **110**, 514–521 (1934)
11. Spira, R.: Zero-free region for $\zeta^{(k)}(s)$. J. Lond. Math. Soc. **40**, 677–682 (1965)
12. Spira, R.: Another zero-free region for $\zeta^{(k)}(s)$. Proc. Am. Math. Soc. **26**(2), 246–247 (1970)
13. Stein, W., et al.: Sage, Open-source Mathematics Software. http://www.sagemath.org (2012)
14. Verma, D.P., Kaur, A.: Zero-free regions of derivatives of Riemann Zeta function. Proc. Indian Acad. Sci. Math. Sci. **91**(3), 217–221 (1982)
15. Yıldırım, C.Y.: Zeros of $\zeta''(s)$ and $\zeta'''(s)$ in $\sigma < 1/2$. Turk. J. Math. **24**(1), 89–108 (2000)

# Chapter 11
# Application of Object Tracking in Video Recordings to the Observation of Mice in the Wild

**Matina Kalcounis-Rueppell, Thomas Parrish, and Sebastian Pauli**

## 11.1 Introduction

Methods for the automating the processing of digital video have been a topic of research since the mid-1980s [1]. These techniques have been used extensively in traffic surveillance and security. In the past decade, automated analysis of video has become increasingly popular in the study of animal behavior, both in the laboratory and in the wild. For example, the individual and social behaviors of fruit flies in a planar arena in a laboratory setting have been quantified using data obtained with computer vision methods [4].

As part of a larger study examining vocal communication among wild deer mice (*Peromyscus* species) [6] infrared video was collected over 131 nights from dusk until dawn. The video was taken from a camera suspended in the tree canopy above the free-living mice on the forest floor. The video was recorded nonstop, regardless of the level of mouse activity. Thus, the volume of video recordings obtained in this study is a challenge to manually process. Computer vision techniques, however, allow us to detect and record the trajectories of moving objects from the video data without human intervention. In the initial phase of the project, mouse trajectories were extracted from short clips of the video recordings with the goal of analyzing the speed of mice [13] and data extracted from the video was validated by a human observer [2]. As the result of this experience we are now able to process the approximately 1,500 h of video and extract biologically meaningful data.

M. Kalcounis-Rueppell
Department of Biology, University of North Carolina Greensboro, Greensboro, NC 27402, USA
e-mail: mckalou@uncg.edu

T. Parrish (✉) • S. Pauli
Department of Mathematics and Statistics, University of North Carolina
Greensboro, Greensboro, NC 27402, USA
e-mail: tmparri2@uncg.edu; s_pauli@uncg.edu

J. Rychtář et al. (eds.), *Topics from the 8th Annual UNCG Regional Mathematics and Statistics Conference*, Springer Proceedings in Mathematics & Statistics 64, DOI 10.1007/978-1-4614-9332-7_11, © Springer Science+Business Media New York 2013

In this paper we report on the methods we used to track the movement of mice in video material and describe how we obtained biologically relevant information from the tracking data, namely measures of mouse activity. The results of our analysis are subject of a forthcoming publication by the authors.

### 11.1.1 Notation

We will use the following notation in our discussion of video and image data. We represent an image as an $m \times n$ matrix $F \in C^{m \times n}$, where $C$ denotes a color space. We denote the $(x, y)$ entry in $F$ by $F_{x,y}$ and refer to it as a picture element, or *pixel*.

Common examples of color spaces are black and white ( $C_0 = \{0, 1\}$ ), grayscale ( $C_g = \{0, \ldots, 255\}$ ), and true color ( $C_t = \{(R, G, B) \mid R, G, B \in \{0, \ldots, 255\}\}$ ). For ease of presentation we will limit most of our discussions to grayscale images and video. It can be easily generalized to other color spaces.

A video $V$ is a sequence of images,

$$V = (F_1, F_2, \ldots, F_n), \tag{11.1}$$

where $n \in \mathbb{N}$ is the number of images in the video. Each image is called a frame, and those frames are displayed at a constant frame rate, which is typically 24, 25, or 30 frames per second.

## 11.2 Foreground Isolation

One of the most fundamental applications of automated video processing is the identification and tracking of moving objects. The most common tracking method is referred to as *blob tracking*. This process involves isolating foreground from background information by means of background subtraction, identifying foreground *connected components*, or collections of adjoined pixels, and tracking those over time.

For our purposes, we consider each pixel of a video image to belong to either the foreground or background. We define the background to be the set of static, or predominantly unchanging pixels, and the foreground to be the set of all other pixels.

The foreground isolation functions return a black and white image $M$ called the foreground mask. A pixel of value $M_{x,y} = 0$, or black, represents a background pixel, and a pixel of value $M_{x,y} = 1$, or white, corresponds to a foreground pixel. We call elements of the foreground objects, and their corresponding foreground mask elements blobs.

## 11.2.1 Background Subtraction

If the pixels corresponding to the background are known, then the foreground can be extracted by taking the absolute *difference* $s(F, B)$ of a frame $F$ and a reference background image $B$, where

$$s : C^{m \times n} \times C^{m \times n} \to C^{m \times n}, \ s(F, B) = G \text{ where } G_{x,y} = \mid F_{x,y} - B_{x,y} \mid . \quad (11.2)$$

Clearly if $s(F, B)_{x,y} = 0$, then $F_{x,y}$ belongs to the background. Because we want to allow some fluctuation in the background pixels a threshold function is used to decide whether a pixel belongs to the foreground or background:

$$t : C_g^{m \times n} \times C_g \to C_0^{m \times n}, \ t(F, c) = G \text{ where } G_{x,y} = \begin{cases} 0 & \text{if } F_{x,y} < c \\ 1 & \text{else} . \end{cases} \quad (11.3)$$

For each frame $F$, if $F' = s(F, B)$, then the foreground mask can be given by $t(F', c)$, where $c$ is typically greater than 200 for grayscale images.

There are various methods for determining the background image, which can be static or updated with every frame, for example:

*First Frame Method.* If the first frame of the video only consists of background, the first frame can be used as the background image. This yields the fastest background subtraction method.

*Average Frame Method.* The average of all frames of the video is used as a background image. This can work even if objects are present in the foreground of all frames, as long as those objects move frequently. Because the entire video must be processed prior to tracking, this method does not allow video processing in real time.

*Running Average of Frames Method.* Using the running (weighted) average of all previous frames as the background image yields better results, particularly when there are frequent subtle changes in lighting. Typically, the background $B$ is initialized to the first frame $F_0$, and after processing each subsequent frame $F$, $B$ is updated to $w_\alpha(F, B)$, where

$$w_\alpha : C^{m \times n} \times C^{m \times n} \to C^{m \times n}, \ (w_\alpha(F, B))_{x,y} = \lfloor \alpha F_{x,y} + (1 - \alpha) B_{x,y} \rfloor \quad (11.4)$$

for some $\alpha \in (0, 1)$.

## 11.2.2 Dilation and Erosion

Often, a foreground pixel is similar in intensity or color to the corresponding background pixel. In this case, the foreground pixel is likely to be improperly classified as a background pixel. This can result in hole within a connected component, or two distinct connected components that represent the same object.

To prevent such errors, a series of morphological operations can be applied namely *dilation* and *erosion*. Dilation increases the size of blobs, merging blobs that represent the same object and removing holes. Erosion reduces the size of blobs and smoothes the edges. These operations are often combined with foreground isolation techniques.

In each operation, the value of a pixel $F_{x,y}$ is set to either the lightest or darkest pixel value in the neighborhood specified by a *kernel*. The kernel can be described as a set of relative coordinates $K \subset Z \times Z$. The neighborhood of $F_{x,y}$ specified by $K$ consists of the pixels with coordinates in $\{(x+i, y+j) \mid (i,j) \in K\}$. The dilation of an image $F$ using the Kernel $K$ is

$$d_K : C^{m \times n} \to C^{m \times n}, \ d_K(F) = G \text{ with } G_{x,y} = \max\{F_{x+i,y+j} \mid (i,j) \in K\}. \tag{11.5}$$

The erosion of $F$ using the Kernel $K$ is

$$c_K : C^{m \times n} \to C^{m \times n}, \ c_K(F) = G \text{ with } G_{x,y} = \min\{F_{x+i,y+j} \mid (i,j) \in K\}. \tag{11.6}$$

It is common to choose a simple kernel, such as $K = \{(i,j) \mid i,j \in \{-1,0,1\}\}$.

Typically, a series of dilation and erosion operations are applied to the foreground mask in the form of *open* and *close* operations, where opening is the dilation of an erosion, and closing is the erosion of a dilation. Both opening and closing will result in blobs very close to their original size.

### 11.2.3 An Advanced Method

More often than not, however, videos of interest will not contain a stationary background. In such cases, it is necessary to seek more intelligent methods of distinguishing foreground pixels from background pixels. The method chosen for our application, developed by Liyuan Li, Weimin Huang, Irene Y.H. Gu, and Qi Tian, uses a Bayes decision rule to classify objects as foreground and background [9]. It is designed to accommodate two types of changes in background state: gradual changes, such as changes in natural lighting, and rapid changes, such as a camera rotation or tree branch movement. Stationary background pixels are classified by their color features, while moving background elements are classified by their color co-occurrence features. The algorithm consists of four steps: detection of state changes, classification of state changes, foreground object identification, and background learning and maintenance. For each frame, the following steps are executed:

1. Generate background model
2. Perform simple background subtraction to remove pixels of insignificant change
3. Classify each remaining pixel as stationary or moving

4. If stationary, compare pixel value with learned color states and use a Bayes rule to determine probability of being foreground
5. If a pixel is classified as moving, compare color co-occurrence, along with color to the set of learned states, and use Bayes rule to determine probability of being foreground
6. Assign pixel to foreground or background accordingly
7. Perform a pair of dilate–erode and erode–dilate operations to remove artifacts and connect blobs
8. Update the set of learned color states and color co-occurrence states
9. Update the reference background image

## 11.3 Component Identification and Labeling

In order to identify specific elements of an image, it is important to identify the connected components, which exist as sets of *neighboring* pixels. In this application, two pixels are considered neighbors if the distance between them is less than or equal to $\sqrt{2}$ pixels.

One way to identify objects is to use component *contours* as the primary identifying feature of each object. An object's contour is its set of edge pixels.

A simple method of identifying and labeling components in an image $F \in C^{m \times n}$ involves generating an associated label image, $L \in \mathbb{N}^{m \times n}$, with each pixel $L_{x,y}$ consisting of the label corresponding to the pixel $F_{x,y}$. An extremely efficient method, proposed by Fu Chang, Chun-Jen Chen, and Chi-Jen Lu can be used for this task [7].

In the algorithm they present, an image $F$ is processed left to right, and top to bottom. When an external contour pixel is encountered, the entire contour is traced and, for each pixel $F_{x,y}$ in the contour, we set $L_{x,y} = l$, where $l \in \mathbb{N}$ is unique to this contour. Once the contour has been traced, foreground pixels inside the contour are also labeled $l$. If an *internal* contour point is reached, the internal contour is again traced, and labeled $l$. When a new external contour pixel is found, it is labeled $l + 1$, and the tracing process repeats. Each set of pixels of the same label is referred to as a blob.

## 11.4 Blob Tracking

In each frame, blobs are labeled by order of detection, making it difficult to ensure a unique label preservation between frames. Because of this, a blob will often have many labels over time, some of which may correspond to labels assigned to other blobs. It is then necessary to check each successive frame and ensure that for any given blob, its label in the current frame corresponds to its label in the previous frame. There are a number of methods to accomplish this. One simple approach is to calculate a set of identifying features, such as size, location, location of centroid, orientation, intensity or color for each blob. After labels are assigned in each frame, the features of each blob are compared to those of every blob in the previous frame that is within a set distance, and labels are re-assigned accordingly. The set of features for each blob can then be output as track information, sorted by blob label.

### 11.4.1   Tracking Data

Because video frames are processed sequentially, blob data generated by the tracker are returned in sequential order. After each frame, the tracker returns data for each blob, consisting of the unique label of the blob (not to be confused with the labels of the components in the frame), its position, its size, and the number of frames the blob has been present. Additional information, such as bounding boxes, histogram information (of use in color video), velocity and acceleration vectors, can also be extracted. However, because it would require inference, rather than direct observation, to generate velocity and acceleration data, introducing uncertainty, these data were not produced. In addition, because the thermal videos are in greyscale, color information was ignored.

## 11.5   Object Tracking in the Mouse Videos

We describe the video material with which we worked, how the tracking was done, and discuss some challenges we encountered and some decisions we needed to make to obtain as much usable data as possible.

The videos were recorded during research where audio, video, and telemetry data were used to analyze the ultrasonic vocalizations of two species of free-living mice, *Peromyscus californicus* and *P. boylii*. The fieldwork took place over 131 nights at the Hastings Natural History Reservation in upper Carmel Valley, California, USA, during the winters of 2008 and 2009. A detailed description of the methods, with example data representing audio, video, and telemetry, can be found in [6].

### 11.5.1   The Mouse Videos

A thermal-imaging camera was suspended by a simple pulley system in the tree canopy approximately 10 m above the ground, allowing continued recording of active mice in the field of view, through the night. The camera used was a Flir Photon 320 with a resolution of 320 by 240 pixels at 30 frames per second in greyscale. The video was recorded with a JVC Everio GZ-MG 555 hard disk camcorder connected to the camera with a composite video cable at an upscaled resolution of 720 by 480 pixels. In the following all pixel measures refer to pixels in the recorded video.

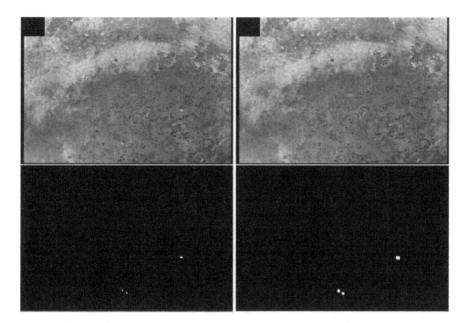

**Fig. 11.1** The four images show a still from an infrared video, background image, the foreground mask, and the foreground mask after dilating twice. The two blobs on the left are partially concealed mice, the blob on the right is another mouse

## 11.5.2  Our Implementation

Previously available animal tracking software was primarily designed for the analysis of animal behavior in a laboratory setting [5, 8], with animals moving in front of a stable background. This specialization makes them less suitable for processing videos of animals in natural environments, where lighting changes and background movement occur frequently. Moreover, many relevant behaviors will be seen in natural environments without a stable background.

For this reason, we wrote a tracking program based on the C++ libraries OpenCV [12] and cvBlob [10], which are freely available under a BSD licence and the LGPL, respectively. OpenCV provided implementations of the algorithms needed for the foreground identification (where we used the advance method described in Sect. 11.2) and the image clean-up steps.

Because of the small size of the mice (about 40 square pixels in the upscaled resolution, 10 square pixels at camera resolution) we use two dilation and no erosion steps in the image cleanup after foreground identification.

The foreground isolation and clean-up steps are illustrated in Fig. 11.1.

The library cvBlob offered the functionality needed for the blob tracking step, including an implementation of the block labeling algorithm described in Sect. 11.3. We found that the simple blob tracking methods implemented in cvBlob were sufficient for our application.

For OpenCV and cvBlob installation instructions, see the web sites given in the references. For an introduction to OpenCV, see the OpenCV book [3]. The functions for the post processing were written using the Python-based computer algebra system Sage. The blob tracking program outputs tracking information in the form of a raw text, which is imported into Sage and processed.

A shell script calling the video processing and post processing was written, allowing several hundred videos to be processed in one batch.

### 11.5.3 Data Filtering

Although the program is able to disregard most noise, some noise may be categorized as legitimate foreground information. However, these false tracks typically have very short durations. For this reason, we have chosen to ignore tracks of extremely short duration, which we classify as tracks less than ten frames long, or one third of a second. It is also the case that a warm wind will occasionally heat up a stationary background element, such as a rock or mouse trap, for a time longer than ten frames. To account for these false tracks, we discard any track for which there is no movement.

### 11.5.4 Blob Classification

Once the tracks are filtered, blobs are categorized based on size and speed. For mice, we calculated an expected size based on known biological size ranges, which we converted to a pixel area based on the dimensions of each focal area. Because these dimensions varied across focal areas, we used a separate range for each area. In addition, we found that bats and birds traveled significantly faster than mice. Any object that traveled faster than three pixels per frame was considered to be a flying vertebrate.

## 11.6  Analysis of Tracks

We used the tracking information in two ways. In the first application, which we refer to as *computer-aided observation*, data were searched for information that targets specific events of interest to human investigators, who then analyzed these events.

In the second application, which we refer to as *automated analysis*, the computer directly computes data, which can then be used for the (statistical) analysis of behavior.

## 11.6.1 Computer-Aided Observation

Computer-aided observation is useful for finding specific events which require qualitative analysis. An example of such an application may be to have the computer extract all times in a video when several objects exist in concurrence. The investigator could then watch the video, in order to determine if the objects (animals) influence each other's behavior.

A script was written to report all times when objects of specific size ranges appear in videos. These size ranges were used for two purposes. We used them to find predators such as cougars (*Puma concolor*), bobcat (*Lynx rufus*), and foxes (*Urocyon cinereoargenteus*), by searching for large blobs, which had an area greater than 500 square pixels. The times when large blobs were present were used as a queues for manual observation, so that these blobs could be classified and behaviors analyzed.

We also returned all times when objects in the expected size range of mice (80–120 square pixels after dilation, depending on focal area) existed for a period of at least 5 s. From this list, we selected a random sample of videos and times and observed the videos. In all cases, we found that the blobs in our expected size ranges corresponded to mice.

## 11.6.2 Automated Analysis

Although computer-aided observation is a valuable tool, it is desirable for the computer to do as much analysis as possible. While the analysis of complex events and interactions is difficult, some data lend itself to easy analysis. Examples of such data include analysis of size distributions, speed of travel, and location preference (i.e., objects do have a tendency to be found in one region more often than another). Our primary application of automated analysis was to analyze levels of mouse activity.

## 11.6.3 Measuring Mouse Activity

Often mice exit and reenter the field of view, or become temporarily masked under dense vegetation. Because of the uncertainty introduced by these events, a decision was made to use only observed data, and to not interpolate missing data. In addition, accurate identification of individuals is difficult due to a lack of identifying features in thermal video. As such, measures of activity that do not require the identification of individual mice were chosen. In this way we avoid introducing unnecessary error.

Assume that a track is active from frame number $m$ to frame number $n$. Let $(x_t, y_t)$ be the position of a blob, at frame number $t$. Because of the high sampling rate of the position of the blob at 30 times per second

$$d = \sum_{t=m}^{n} \sqrt{(x_t - x_{t-1})^2 + (y_t - y_{t-1})^2} \qquad (11.7)$$

is a good approximation of the length of the track.

To measure the activity of mice on a given night, we use two values:

1. the total observed distance $D$ travelled by all mice throughout the night; that is, the sum of the lengths of all observed tracks and
2. the average speed $S$ of all mice throughout the night; that is, $S = D/T$ where $T$ is the sum of the lengths in time of all observed tracks.

These measures make it possible to investigate the change in mouse activity under various biotic and abiotic environmental influences. This investigation is subject of a forthcoming publication of the authors.

## 11.7   Conclusion

Automated tracking is remarkably useful. With a limited understanding of computer vision techniques and moderate computer programming experience, it is possible to construct an automated video processing program suitable for analyzing some types of animal behavior. The results obtained from these types of programs, e.g. tracking information, help us to answer numerous biological questions and save researchers a great deal of time. Useful information can often be obtained from even poor quality video.

Some caveats exist, however. For example, it is difficult to distinguish amongst individuals in grayscale video. Also, it is difficult to extract accurate tracking data from videos containing large amounts of background movement, which is often a result of wind when a camera is setup with a hanging-pulley system. An easy solution would be to anchor the camera in such a way so that swaying in windy conditions would be prevented.

We believe that automated video processing provides a meaningful alternative to traditional methods of studying animal behavior, especially that of a nocturnal, secretive species. Past behavioral studies have resorted to methods such as trapping [11], sand transects [14], or test arenas [15]. With proper setup, remotely recorded video, along with automated video processing techniques, can provide information not traditionally available. This information includes data such as speed, distance traveled, frequency of travel, and number of animals in a given space at a given time. This type of information in a natural setting provides crucial information to better understand the evolution and maintenance of behaviors in natural contexts.

The use of thermal imaging allows for the collection of these types of data on secretive and nocturnal rodents. Moreover, automated video processing presents a means to efficiently analyze the behaviors present in such videos, although it is equally capable of analyzing behavior in traditional video.

**Acknowledgments** The work on this project was supported by National Science Foundation (Grants IOB-0641530, IOB-1132419, DMS-0850465 and DBI-0926288). We would like to thank the Office of Undergraduate Research of UNCG and in particular its directors Mary Crowe and Jan Rychtar. We thank Shan Suthaharan for bringing the group for the initial research project together and David Schuchart for the tracking program that he wrote for the initial project [13]. Thanks also go to Christian Bankester for his work on video analysis [2], Caitlin Bailey, Luis Hernandez, all the students who worked in the field collecting data, and the Hastings Natural History Reserve for all of their support of our field work.

# References

1. Andersson, R.: Real-time gray-scale video processing using a moment-generating chip. IEEE J. Robotic Autom. **1**, 79–85 (1985)
2. Bankester, C., Pauli, S., Kalcounis-Rueppell, M.: Automated processing of large amounts of thermal video data from free-living nocturnal rodents. http://www.uncg.edu/mat/faculty/pauli/mouse/reu2011.html (2011)
3. Bradski, G., Kaehler, A.: Learning OpenCV: Computer Vision with the OpenCV Library. O'Reilly Media, Sebastopol CA, USA (2008)
4. Branson, K., Robie, A.A., Bender, J., Perona, P., Dickinson, M.H.: High-throughput ethomics in large groups of Drosophila. Nat. Methods **6**, 451– 457 (2009). http://dx.doi.org/10.1038/nmeth.1328
5. Branson, K., et al.: CTRAX – the caltech multiple walking fly tracker. http://ctrax.sourceforge.net. Accessed 2012
6. Briggs, J.R., Kalcounis-Rueppell, M.C.: Similar acoustic structure and behavioural context of vocalizations produced by male and female California mice in the wild. Anim. Behav. **82**, 1263–1273 (2011). http://www.sciencedirect.com/science/article/pii/S0003347211003836
7. Chang, F., Chen, C.-J., Lu, C.-J.: A linear-time component-labeling algorithm using contour tracing technique. Comput. Vis. Image Und. **93**(2), 206–220 (2004)
8. EthoVision XT. http://www.noldus.com/animal-behavior-research/products/ethovision-xt. Accessed 2012
9. Li, L., Huang, W., Gu, I.Y.H., Tian, Q.: Foreground object detection from videos containing complex background. ACM MM (2003)
10. Liñán, C.C.: cvBlob – Blob library for OpenCV. http://cvblob.googlecode.com. Accessed 2012
11. Marten, G.G.: Time patterns of peromyscus activity and their correlations with weather. J. Mammal. **54**(1), 169–188 (1973). http://www.jstor.org/discover/10.2307/1378878?uid=2&uid=4&sid=21101483719073
12. OpenCV – Open Source Computer Vision. http://opencv.willowgarage.com. Accessed 2012
13. Schuchart, D., Pauli, S., Suthaharan, S., Kalcounis-Rueppell, M.: Measuring behaviors of Peromyscus mice from remotely recorded thermal video using a blob tracking algorithm. http://www.uncg.edu/mat/faculty/pauli/mouse/mathbio2010.html (2011)
14. Vickery, W.L., Bider, J.R.: The influence of weather on rodent activity. J. Mammal. **62**(1), 140–145 (1981). http://www.jstor.org/discover/10.2307/1380484?uid=2&uid=4&sid=21101483719073
15. Wolfe, J.L., Tan Summerlin, C.: The influence of lunar light on nocturnal activity of the old-field mouse. Anim. Behav. **37**(Part 3), 410–414 (1989). http://www.sciencedirect.com/science/article/pii/0003347289900882

# Chapter 12
# The Card Collector Problem

Anda Gadidov and Michael Thomas

## 12.1  Introduction

Suppose there are $m$ different cards to complete a certain collection, such as baseball cards or McDonald's Monopoly game pieces. Card of type $i$ occurs independently of the other ones with probability $p_i \geq 0$, $\sum_{i=1}^{m} p_i = 1$. The question is to find the probability of getting a complete collection of cards and the expected number of cards that have to be purchased in order to complete the collection. Assuming equal probabilities the problem was first mentioned in 1709 by DeMoivre in his collection of 26 problems related to games of chance titled *De Mensura Sortis, deu, de Probabilitate Eventuum in Ludis a Casu Fortuito Pendentibus*. In 1938 Kendall and Smith [3] mentioned the problem in relation to checking the randomness of their sampling numbers, Feller [1] presented the question as a type of urn problem, Flajolet et al. [2] used symbolic methods in combinatorial analysis to analyze several related allocation problems. The coupon collector problem is often mentioned in occupancy problems in which balls are thrown independently at a finite or infinite series of boxes. In this context the problem found numerous applications in species sampling problems in ecology, and also in database query optimization.

Let $X$ be the number of cards that need to be purchased in order to complete a collection. Using the inclusion–exclusion principle we derive the probability distribution of $X$ and we compute its expected value. In particular we obtain an interesting identity for the equally likely case. Following ideas of Nakata [4] we show that the minimum expected value is attained in the equally likely case.

A. Gadidov (✉) • M. Thomas
Kennesaw State University, 1000 Chastain Rd. #1601, Kennesaw, GA 30144, USA
e-mail: agadidov@kennesaw.edu; mthom130@students.kennesaw.edu

J. Rychtář et al. (eds.), *Topics from the 8th Annual UNCG Regional Mathematics and Statistics Conference*, Springer Proceedings in Mathematics & Statistics 64, DOI 10.1007/978-1-4614-9332-7_12, © Springer Science+Business Media New York 2013

## 12.2  Results

We introduce the following notation: for a subset $J \subset \{1, 2, \ldots, m\}$ denote by $|J|$ the cardinality of $J$. Also, for $i \in \{1, 2, \ldots, m\}$, $J_i$ will be used to denote a subset of $\{1, 2, \ldots, m\} \setminus \{i\}$. Let $X_i$ denote the number of cards of type $i$ and $A_{i,n}$ the event that card of type $i$ was the last added to the collection when $n$ cards were needed to complete the collection. Define

$$X_J := \sum_{i \in J} X_i, \quad P_J := \sum_{i \in J} p_i. \tag{12.1}$$

**Proposition 1.** *For $n \geq m$ we have:*

$$P(X = n) = \sum_{r=0}^{m-2} \sum_{|J|=m-1-r} (-1)^r (1 - P_J) P_J^{n-1}. \tag{12.2}$$

*Proof.* For $m = 2$ denote by $p$ and $1 - p$ the probability of getting the first and second card, respectively. For $n \geq 2$ we have

$$P(X = n) = p^{n-1}(1 - p) + (1 - p)^{n-1} p. \tag{12.3}$$

For $m = 3$ the collection can be completed in $n \geq 3$ cards by getting the first, second or third card last. Therefore we have

$$P(X = n) = \sum_{i=1}^{3} P(A_{i,n}). \tag{12.4}$$

Let us look at

$$P(A_{1,n}) = p_1 P(X_2 > 0, X_3 > 0, X_2 + X_3 = n - 1)$$
$$= p_1 P(X_2 + X_3 = n - 1) - p_1 P(X_2 = 0, X_3 = n - 1)$$
$$- p_1 P(X_2 = n - 1, X_3 = 0)$$
$$= p_1 \big((p_2 + p_3)^{n-1} - p_2^{n-1} - p_3^{n-1}\big). \tag{12.5}$$

Using Eq. (12.5) for all $A_{i,n}, i = 1, 2, 3$, Eq. (12.4) becomes

$$P(X = n) = p_1 \big((p_2 + p_3)^{n-1} - p_2^{n-1} - p_3^{n-1}\big)$$
$$+ p_2 \big((p_1 + p_3)^{n-1} - p_1^{n-1} - p_3^{n-1}\big)$$
$$+ p_3 \big((p_1 + p_2)^{n-1} - p_1^{n-1} - p_2^{n-1}\big).$$

By regrouping the terms we can rewrite $P(X = n)$ in a format that will be useful when we treat the general case:

$$P(X = n) = \sum_{i=1}^{3} \left( p_i (1 - p_i)^{n-1} - p_i^{n-1}(1 - p_i) \right). \tag{12.6}$$

For general $m$ we proceed as in the case $m = 3$. For $n \geq m$ we write

$$P(X = n) = \sum_{i=1}^{m} P(A_{i,n}). \tag{12.7}$$

For $i = 1, \ldots, m$, using the inclusion–exclusion formula as in the case $m = 3$ we obtain:

$$P(A_i, n) = p_i P(X = n - 1, X_j > 0, j \neq i, X_i = 0)$$

$$= \sum_{r=0}^{m-2} \sum_{|J_i|=m-1-r} (-1)^r p_i P(X_{J_i} = n - 1)$$

$$= \sum_{r=0}^{m-2} \sum_{|J_i|=m-1-r} (-1)^r p_i P_{J_i}^{n-1}.$$

By grouping all terms that contain the same $P_J^{n-1}$ for some $J \subset \{1, 2, \ldots, m\}, |J| = m - 1 - r$ when summing over $i = 1, 2, \ldots, m$ we obtain:

$$P(X = n) = \sum_{i=1}^{m} \sum_{r=0}^{m-2} \sum_{|J_i|=m-1-r} (-1)^r p_i P_{J_i}^{n-1}$$

$$= \sum_{r=0}^{m-2} \sum_{|J|=m-1-r} (-1)^r (1 - P_J) P_J^{n-1}.$$

$\square$

**Corollary 1.** *In particular, if cards are equally likely*

$$P(X = n) = \sum_{r=0}^{m-2} \binom{m}{m-1-r} \frac{1+r}{m} \left( \frac{m-1-r}{m} \right)^{n-1}. \tag{12.8}$$

*Proof.* If cards are equally likely $p_i = 1/m$ for all $i = 1, \ldots, m$ and since for each $r = 0, \ldots, m - 1$ there are $\binom{m}{m-1-r}$ subsets $J \subset \{1, \ldots, m\}, |J| = m - 1 - r$, the result follows from (12.2). $\square$

The next Lemma will be used in computing the expected number of cards needed to complete the collection.

**Lemma 1.** *Let $0 < p < 1$ and $m \geq 2$ integer. Then*

$$\sum_{n=m}^{\infty} np^{n-1} = \frac{mp^{m-1} - (m-1)p^m}{(1-p)^2}. \tag{12.9}$$

*Proof.* Using convergence of geometric series for $0 < p < 1$, we have

$$\sum_{n=m}^{\infty} p^n = \frac{p^m}{1-p}. \tag{12.10}$$

Now, the series in (12.10) is uniformly convergent on $0 < p < 1$, therefore we may differentiate term by term to obtain:

$$\sum_{n=m}^{\infty} np^{n-1} = \frac{mp^{m-1} - (m-1)p^m}{(1-p)^2}. \tag{12.11}$$

$\square$

**Proposition 2.** *Let $X$ be the number of cards needed to complete a collection. Then*

$$E(X) = \sum_{r=1}^{m-1} \sum_{|J|=r} (-1)^{r-1} \frac{m(1-P_J)^{m-1} - (m-1)(1-P_J)^m}{P_J}. \tag{12.12}$$

*Proof.* Using the probability distribution of $X$ given in (12.2) together with Lemma 1 we obtain:

$$E(X) = \sum_{n=m}^{\infty} nP(X = n)$$

$$= \sum_{n=m}^{\infty} \sum_{r=0}^{m-2} \sum_{|J|=m-1-r} (-1)^r n(1 - P_J)P_J^{n-1}$$

$$= \sum_{r=0}^{m-2} \sum_{|J|=m-1-r} (-1)^r \frac{mP_{J_i}^{m-1} - (m-1)P_J^m}{(1-P_J)}$$

$$= \sum_{r=1}^{m-1} \sum_{|J|=r} (-1)^{r-1} \frac{m(1-P_J)^{m-1} - (m-1)(1-P_J)^m}{P_J}.$$

$\square$

**Corollary 2.** *For every integer $m \geq 2$ the following holds:*

$$\sum_{r=1}^{m-1} \binom{m}{r} (-1)^{r-1} \frac{m\left(1 - \frac{r}{m}\right)^{m-1} - (m-1)\left(1 - \frac{r}{m}\right)^m}{r/m} = m\left(1 + \frac{1}{2} + \frac{1}{3} + \cdots + \frac{1}{m}\right).$$

(12.13)

*Proof.* For the particular case of equally likely cards $p_i = 1/m, i = 1, \ldots, m$ and since there are $\binom{m}{r}$ subsets $J : |J| = r$, the expected value in (12.12) becomes:

$$E(X) = \sum_{r=1}^{m-1} \binom{m}{r} (-1)^{r-1} \frac{m\left(1 - \frac{r}{m}\right)^{m-1} - (m-1)\left(1 - \frac{r}{m}\right)^m}{r/m}.$$

From [5], the expected number of cards needed to complete a collection under the equally likely assignment is $m\left(1 + \frac{1}{2} + \frac{1}{3} + \cdots + \frac{1}{m}\right)$, therefore the result follows. □

Following Nakata's approach in [4], we show that the equally likely case has the smallest expected value among all possible probability distributions on the set of $m$ cards. Denote by $p_{[j]}$ the j-th largest value of the vector $(p_1, \ldots, p_m)$, so $p_{[1]} \geq p_{[2]} \geq \cdots \geq p_{[m]}$.

**Definition 1.** For two probability vectors $\mathbf{p} = (p_1, \ldots, p_m), \mathbf{q} = (q_1, \ldots, q_m)$ we say that $\mathbf{p}$ is **majorized** by $\mathbf{q}$, denoted $\mathbf{p} \prec \mathbf{q}$ if

$$\sum_{j=1}^{k} p_{[j]} \leq \sum_{j=1}^{k} q_{[j]}, 1 \leq k \leq m - 1.$$

(12.14)

For example, $\mathbf{p} = (.32, .28, .40) \prec \mathbf{q} = (.35, .23, .42)$.

**Theorem 1.** *The equally likely probability assignment is majorized by any other probability assignment:*

$$\mathbf{p} = \left(\frac{1}{m}, \ldots, \frac{1}{m}\right) \prec \mathbf{q} \text{ for any } \mathbf{q}$$

(12.15)

*Proof.* Let $\mathbf{q}$ be an arbitrary probability distribution vector. If $q_{[1]} < 1/m$, then $q_{[j]} \leq q_{[1]} < 1/m$, for all $j = 1, \ldots, m$ and it follows that $\sum_{j=1}^{m} q_j < 1$. Therefore $q_{[1]} \geq 1/m$.

Let now $1 < k < m$ be the smallest index for which $\sum_{j=1}^{k} q_{[j]} < \frac{k}{m}$. Since $\sum_{j=1}^{k-1} q_{[j]} \geq \frac{k-1}{m}$, it follows that $q_{[i]} < 1/m$ for all $i = k, \ldots, m$. But then again we obtain that $\sum_{j=1}^{m} q_{[j]} < 1$, contradiction! The conclusion follows. □

**Definition 2.** A function $f(\mathbf{p})$ which is symmetric in $p_1, p_2, \ldots, p_m$ is **Schur convex** if $\mathbf{p} \prec \mathbf{q} \Rightarrow f(\mathbf{p}) \leq f(\mathbf{q})$.

**Definition 3.** For $X, Y$ random variables we say that $X$ is **stochastically smaller** than $Y$, denoted $X \leq_S Y$ if for all $a \in R, P(X > a) \leq P(Y > a)$.

Denote by $X_\mathbf{p}$ the number of cards that need to be purchased to complete the collection when the probability distribution is given by the vector $\mathbf{p}$. The next result stated without proof can be found in [4].

**Theorem 2 (Nakata).** $P(X_\mathbf{p} > n)$ *is a Schur convex function of* $\mathbf{p}$.

**Theorem 3.** *The expected value of the number of cards needed to complete the collection is minimum when the cards are equally likely.*

*Proof.* Let $\mathbf{p_0}$ denote the probability vector corresponding to the uniform distribution on $\{1, \ldots, m\}$. It is known that for a discrete random variable taking positive values $E(X) = \sum_{n=0}^{\infty} P(X > n)$. Then for any other probability vector $\mathbf{q}$, using Theorems 1 and 2 we have:

$$E(X_{\mathbf{p_0}}) \leq E(X_\mathbf{q}), \qquad (12.16)$$

therefore the result follows.                                                                                                     □

## 12.3  Final Remarks

Using generating functions the authors in [2] derive an expression for the expected time to obtain a partial collection of $j$ cards, and in particular a complete collection [Eqs. (14a) and (14b)]. Although we use the same meaning for the probability $P_J$, our expression for the expected value to complete the collection, (12.12), is different. As a consequence of our approach we obtain the nontrivial identity:

$$\sum_{q=0}^{m-1} (-1)^{m-1-q} \sum_{|J|=q} \frac{1}{1-P_J} = \sum_{r=1}^{m-1} \sum_{|J|=r} (-1)^{r-1} \frac{m(1-P_J)^{m-1}-(m-1)(1-P_J)^m}{P_J}$$

$$(12.17)$$

In particular, for the equally likely case we have the identity mentioned in Corollary 2.

**Acknowledgements**  Presentation of the results at the UNCG Regional Mathematics and Statistics Conference was made possible by a Mentor Protégé grant offered by the College of Sciences and Mathematics at Kennesaw State University.
We would like to thank the referee for carefully reading the paper and making some useful suggestions.

## References

1. Feller, W.: An Introduction to Probability Theory and Its Applications, vol. 1, 3rd edn. Wiley, New York (1968)

2. Flajolet, P., Gardy, D., Thimonier, L.: Birthday paradox, coupon collectors, caching algorithms and self-organizing search. Discr. Appl. Math. **39**, 207–229 (1992)
3. Kendall, M.G., Babington Smith, B.: Randomness and random sampling numbers. J. R. Stat. Soc. **101**, 147–166 (1938)
4. Nakata, T.: Card collector's problem with unequal probabilities. Available online http://www.fukuoka-edu.ac.jp/~nakata/papers/coumaj.pdf (2008)
5. Von Schelling, H.: Coupon collecting for unequal probabilities. Am. Math. Month. **61**, 306–311 (1954)

# Chapter 13
# The Effect of Information on Payoff in Kleptoparasitic Interactions

Mark Broom, Jan Rychtář, and David G. Sykes

## 13.1 Introduction

Kleptoparasitism, the stealing or attempted stealing of resources (usually food), is a very common behavior practiced by a very diverse collection of species such as insects [14], fish [12], birds [16–18], and mammals [15]. For a recent review paper with complete classification and numerous examples, see [13].

The strategies associated with stealing interactions can vary; for instance, sometimes resources are promptly forfeited while in other cases the individuals defend the resources vigorously and even engage in fights.

The effect of variation in resource value on fighting behavior was investigated in detail in [11], who used a simulation model to investigate a situation where a resource owner possesses information about the (subjective) value of a resource that an individual attempting to steal it may or may not have, using a sequential assessment game. Their model predictions included that the resource owner's probability of victory would increase with increasing resource value, based partly upon the extra knowledge that the owner had (but see [5]), and that costs and contest duration will also increase with resource value.

However, in most models, see, for example, [2, 4, 6] and references therein, the individuals value the resource equally even when the resources can differ in value

M. Broom (✉)
Department of Mathematics, City University London, Northampton Square,
London EC1V 0HB, UK
e-mail: mark.broom@city.ac.uk

J. Rychtář • D.G. Sykes
Department of Mathematics and Statistics, The University of North Carolina
at Greensboro, Greensboro, NC 27402, USA
e-mail: rychtar@uncg.edu; dgsykes@uncg.edu

J. Rychtář et al. (eds.), *Topics from the 8th Annual UNCG Regional Mathematics and Statistics Conference*, Springer Proceedings in Mathematics & Statistics 64, DOI 10.1007/978-1-4614-9332-7_13, © Springer Science+Business Media New York 2013

such as in the situations investigated in [3, 5]. The variation in value can be caused by external factors such as the size of the food item; however, it can be caused by an internal state (such as hunger) of the individuals [11].

As soon as there is a difference between individuals in resource valuation, several informational situations arise. Firstly, when individuals are aware of their own as well as their opponent's valuation. Secondly, when individuals are aware only of their own valuation. Thirdly, when individuals are not aware even of their own valuation.

A common way to model kleptoparasitic interactions is the so-called producer-scrounger game developed in [1]. A number of variants of this model have been developed to consider different circumstances and assumptions (see, for example, [8–10, 19]). One advantage of this type of model is that analysis is relatively straightforward, so that clear predictions can be made. Here, we consider a scenario where one individual, a producer, possesses a valuable resource when another individual, a scrounger, comes along and may attempt to steal it.

## 13.2 The Model

We model the situation of a scrounger discovering a producer with a resource as a sequential game in extensive form as shown in Fig. 13.1. If the scrounger makes such a stealing attempt, then the producer can either give up the resource without any conflict or defend it. The conflict cost is $c$ and the producer wins the conflict (and can keep the resource) with probability $a$.

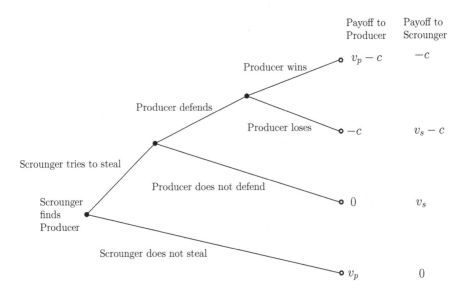

**Fig. 13.1** Scheme and payoffs of the game

Let us denote the value for the scrounger as $v_s$, and the value for the producer as $v_p$. We assume that the distributions of $v_s$ and $v_p$ are the same. The game and the payoffs from different scenarios are shown in Fig. 13.1.

## 13.3   Analysis

We will analyze the game using backward induction, see, for example, [7, p. 187].

### 13.3.1   Full Information Case

Here we assume that individuals know the resource values for themselves as well as for their opponents. Assume that the scrounger attempts to steal. The producer has to decide whether to defend or not. If the producer does not defend, the payoff will be 0. If the producer defends, individuals will fight and the producer will lose it with probability $1 - a$. Hence, the producer's expected payoff when defending is $av_p - c$. Consequently, the producer should defend only if $0 < av_p - c$ which is equivalent to

$$\frac{c}{a} < v_p. \tag{13.1}$$

Note that the producer does not need to know the value of the resource for the scrounger. All that is relevant to the producer is the fact that the scrounger attempted to steal and then the producer can evaluate the payoffs to itself.

Now, we will investigate the options for the scrounger, assuming it knows $v_p$. If the scrounger does not attempt to steal, the payoff will be 0. If (13.1) does not hold, then the producer will not defend against a stealing attempt and thus the scrounger should attempt to steal to get a payoff $v_s > 0$. If (13.1) holds, then the producer will defend against the stealing attempt. Hence, if the scrounger attacks, it will lose with probability $a$ (and get a payoff $-c$) and win with probability $1 - a$ (and get a payoff $v_s - c$). The expected payoff is thus $(1 - a)v_s - c$. Hence, the scrounger should attack if

$$(1 - a)v_s - c > 0 \tag{13.2}$$

which is equivalent to

$$\frac{c}{1 - a} < v_s. \tag{13.3}$$

There are thus three distinct behavioral patterns as presented in Table 13.1 and Fig. 13.2.

**Table 13.1** Summary of the results

| Behavioral outcome | | Condition for full information | | Condition for partial information | | Condition for no information |
|---|---|---|---|---|---|---|
| Scrounger steals | Producer defends | $v_s$ | $v_p$ | $v_s$ | $v_p$ | $E[v]$ |
| Yes | No | any | $v_p < \frac{c}{a}$ | $v_s > \frac{c\pi}{1-a\pi}$ | $v_p < \frac{c}{a}$ | $E[v] < \frac{c}{a}$ |
| No | Yes | $v_s < \frac{c}{1-a}$ | $v_p > \frac{c}{a}$ | $v_s < \frac{c\pi}{1-a\pi}$ | any | $E[v] < \frac{c}{1-a}$ $\quad E[v] > \frac{c}{a}$ |
| Yes | Yes | $v_s > \frac{c}{1-a}$ | $v_p > \frac{c}{a}$ | $v_s > \frac{c\pi}{1-a\pi}$ | $v_p > \frac{c}{a}$ | $E[v] > \frac{c}{1-a}$ $\quad E[v] > \frac{c}{a}$ |

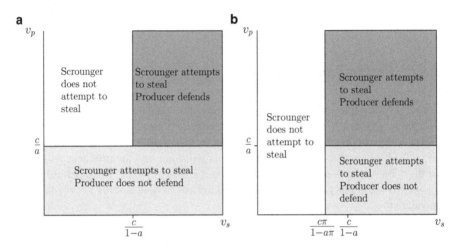

**Fig. 13.2** Behavioral outcomes of the game for the same parameter values $c$ and $a$ but different information cases. (**a**) Full information case, (**b**) Partial information case. We note that $\pi$ actually depends on $c$, and if $c$ is large enough, $\pi = 0$ i.e., the white region can disappear

### 13.3.2 Partial Information Case

Now, assume that the scrounger knows the value $v_s$ and the distribution of $v_p$ (which is assumed to be the same as distribution of $v_s$; in particular, it does not depend on the value of $v_s$), but does not know the exact value of $v_p$. Consequently, the scrounger does not know for sure whether the producer will defend. However, it is still true that the producer will defend if $\frac{c}{a} < v_p$. From the scrounger's perspective, the producer will thus defend with a probability $\pi = \text{Prob}\left(\frac{c}{a} < v_p\right)$. If the producer does not defend, the payoff to the scrounger will be $v_s$. If the producer defends, the payoff to the scrounger will be $(1 - a)v_s - c$. Hence, if the scrounger attempts to steal, his payoff will be

$$(1 - \pi)v_s + \pi\big((1 - a)v_s - c\big) = v_s(1 - a\pi) - c\pi. \tag{13.4}$$

If the scrounger does not attempt to steal, its payoff will be 0. Hence, the scrounger should attempt to steal if

$$v_s > \frac{c\pi}{1 - a\pi}. \tag{13.5}$$

There are thus three behavioral patterns as presented in Table 13.1 and also in Fig. 13.2.

### 13.3.3   No Information Case

The analysis in the no information case is actually very similar to the full information case. The only difference is that the individuals do not know the exact value of the resource, but they do know the expected values, $E[v_p]$ and $E[v_s]$. Since we assume that the distributions of $v_p$ and $v_s$ are the same, we have $E[v_p] = E[v_s]$ and we will denote it just by $E[v]$. There are thus three distinct behavioral patterns as presented in Table 13.1.

## 13.4   Comparison Between Different Information Cases

The illustrative comparison is shown in Fig. 13.3 in the case where the values $v_s$ and $v_p$ have uniform distribution between $v_{\min}, v_{\max}$ and are independent.

### 13.4.1   Comparison Between the Full and Partial Information Cases

Since the function $f(x) = \frac{cx}{1-ax}$ is increasing in $x$ and $0 \le \pi \le 1$, we get that

$$\frac{c}{1 - a} \ge \frac{c\pi}{1 - a\pi} \tag{13.6}$$

with equality only if $\pi = 1$.

For now, let us consider that $\pi \in (0, 1)$. It follows from (13.3), (13.5), and (13.6) that when $v_s > \frac{c}{1-a}$, the scrounger steals regardless of $v_p$ and thus the scrounger's expected payoff (given any distribution of $v_p$ for the producer) is the same in the full information and partial information cases.

On the other hand, if $v_s < \frac{c\pi}{1-a\pi}$ (which is possible only if $\pi > 0$), then the scrounger does not steal in the partial information case, leaving it with the payoff 0. However, if the scrounger knew $v_p$, it would steal if

$$v_p < \frac{c}{a} \tag{13.7}$$

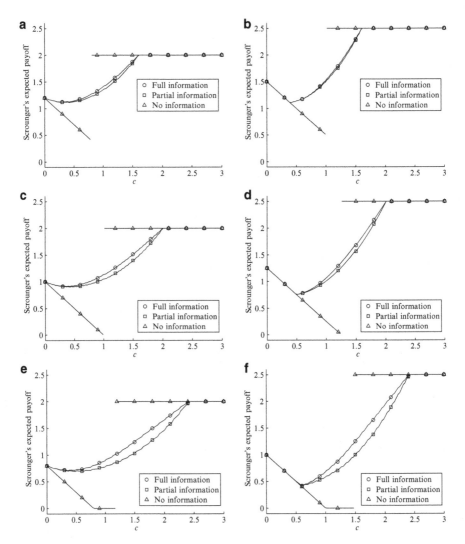

**Fig. 13.3** The Scrounger's payoffs for varying cost of the fight $c$, different distribution of $v$ and different values of $a$ in three different information scenarios. (**a**) $a = 0.4$, $v$ uniformly distributed in $[0, 4]$, (**b**) $a = 0.4$, $v$ uniformly distributed in $[1, 4]$, (**c**) $a = 0.5$, $v$ uniformly distributed in $[0, 4]$, (**d**) $a = 0.5$, $v$ uniformly distributed in $[1, 4]$, (**e**) $a = 0.6$, $v$ uniformly distributed in $[0, 4]$, (**f**) $a = 0.6$, $v$ uniformly distributed in $[1, 4]$

and in those cases the scrounger's payoff would be $v_s$. When $\pi < 1$, the distribution of $v_p$ is such that (13.7) is satisfied with positive probability $1 - \pi$, and thus the expected payoff to the scrounger in the full information case is positive (i.e., larger than the expected payoff in the partial information case, which is 0).

It remains to investigate the values $v_s$ such that $\frac{c\pi}{1-a\pi} < v_s < \frac{c}{1-a}$. For such $v_s$, if $v_p$ is such that (13.7) holds, then the payoff to the scrounger is $v_s$ and the payoff in the full information and partial information cases are the same. However, if $v_p$ is such that (13.7) does not hold, then, by (13.3), the expected payoff to the stealing scrounger is negative.

Hence, overall, the expected payoff for the scrounger (given the distribution of $v_p$) in the full information case is larger than in the partial information case. One can also see that as $a$ increases, the advantage of the full information case gets larger.

It remains to investigate the cases of $\pi = 0$ and $\pi = 1$. It turns out that in such cases, the expected payoffs for the scrounger are the same. If $\pi = 0$, then $c$ is always larger than $av_p$ and so the Scrounger always steals and the producer never defends. If $\pi = 1$, then $c$ is always smaller than $av_p$, i.e., the producer always defends and the scrounger behaves the same way in both cases.

### 13.4.2   Comparison Between the No Information Case and the other Cases

Let $c_0 = \inf\left\{c; Prob\left(\frac{c}{a} < v_p\right) = 0\right\}$ and $c_1 = \sup\left\{c; Prob\left(\frac{c}{a} < v_p\right) = 1\right\}$. When $c > c_0$, then $\pi = 0$ and also $\frac{c}{a} > E[v]$. Hence, in any scenario, the scrounger steals and the producer does not defend. Consequently, the expected payoff to the scrounger is the same in all information cases.

When $aE[v] < c < c_0$, the no information case is better for the scrounger than the full information case (which is better than the partial information case). Indeed, in the no information case, the scrounger attempts to steal and the producer always gives up, leaving the scrounger with the expected payoff $v_s$ which it cannot get for any other scenario (since now $\pi > 0$ and hence there is a positive probability of having $v_p > \frac{c}{a}$).

When $(1 - a)E[v] < c < aE[v]$, the scrounger does not attempt to steal, getting a payoff of 0. This is worse for the scrounger than in the partial information case (the scrounger attempts to steal there for some values, sometimes receiving a free resource, and still gets a positive payoff even when the producer defends) which is worse than in the full information case.

When $c_1 < c < \min\{a, (1 - a)\} E[v]$, then in the no information case, the scrounger always attempts to steal and the producer always defends. This is worse for the scrounger than in the partial information case (which is worse than the full information case) since there are items that are not worth fighting for.

When $c < c_1$, then the expected payoffs in all information cases are the same, since the Scrounger always steals and the Producer always defends.

### *13.4.3  Summary*

The amount of information available has no effect on the payoff to the scrounger when the cost of the fight is relatively small (i.e., when $c < c_1$ so that then it is beneficial to fight for any item under any informational situation) or relatively large (i.e., when $c > c_0$ so that for the producer it is not beneficial to fight for any item under any informational situation). For intermediate costs $c \in (c_1, c_0)$, having full information is better than having only partial information. Moreover, if $c < aE[v]$, then the no information case yields even lower payoffs; and when $c > aE[v]$, then the no information case yields the largest expected payoff.

It is clear that the variance of the resource values has a strong influence on our results. If this variance is small, then $c_0$ and $c_1$ will be close together and the intermediate region where behavior differs between the cases is small. Note that if the variance is actually zero, then there is no useful information to be had and the three cases are identical. For large variance, the intermediate region may account for all plausible cases, and the models will yield significantly different results.

## 13.5  Discussion

In this paper we investigated the effect of information on the payoffs of a producer-scrounger game. One would be tempted to argue that having more information would yield larger payoffs and this was indeed the case for a scrounger having full information versus one with only partial information in the model described by this paper; and, for some parameter values, also the case of no information versus full or partial information case.

However, having more information is not always better. The no information case, where an individual does not know the real value of the resource, is for some parameter values the best case for the scrounger. Yet, let us point out that although this was called the no information case, the scrounger has in fact a very valuable piece of information—the scrounger knows that the producer does not know the real value either, and consequently knows whether it will fight a stealing attempt.

We note that the fact that knowing less is sometimes better has already been observed before. In [5], the authors investigate a scenario in which the value of the resource is the same for both the producer and the scrounger, but nevertheless the resource value is variable and either both the producer and the scrounger know the value, or only the producer knows it. When the scrounger knows the value of the resource, its expected payoffs are lower than when he does not know it. Also, in [7, p. 364], the authors discuss a Producer–Scrounger game that is similar to the one described here, yet again, knowing seemingly less yields larger payoffs for the scrounger.

We also note that in Fig. 13.3 we have assumed that the values of $v_s$ and $v_p$ are independent. The relationship between the two is particularly important in the partial information case, where knowledge of $v_s$ may provide the scrounger with information about $v_p$, and so affect $\pi$. Independence of resource valuation is actually one extreme of a spectrum, the other end of which is complete coincidence of the two. The former is more plausible when the valuation is based on hunger; then at least in the first approximation, the fact that one individual is hungry does not give any new information about its opponent, so that the assumption of independence is reasonable in this case. However, it is also true that if one individual is hungry, then it may be largely because there is not much food around and the same will be true for its opponent. Thus the correlation between the resource values may be important. In this case the latter is more plausible, and this will also be the case if food items vary in size.

Finally, the variance of the resource value will also have a significant effect on our results. For low variance the models mainly coincide, but for high variance their predictions can be very different. It is the variability of the resource value which makes the possession or lack of information important, and the combination of variability in the value of the resource and the availability of information which makes this model an interesting one to study.

**Acknowledgments**  The research was supported by NSF grants DMS-0850465 and DBI-0926288, Simons Foundation grant 245400 and UNCG Undergraduate Research Award in Mathematics and Statistics.

# References

1. Barnard, C.J., Sibly, R.M.: Producers and scroungers: a general model and its application to captive flocks of house sparrows. Anim. Behav. **29**(2), 543–550 (1981)
2. Broom, M., Luther, R.M., Ruxton, G.D.: Resistance is useless? Extensions to the game theory of kleptoparasitism. Bull. Math. Biol. **66**(6), 1645–1658 (2004)
3. Broom, M., Ruxton, G.D.: Evolutionarily stable kleptoparasitism: consequences of different prey types. Behav. Ecol. **14**(1), 23 (2003)
4. Broom, M., Rychtář, J.: The evolution of a kleptoparasitic system under adaptive dynamics. J. Math. Biol. **54**(2), 151–177 (2007)
5. Broom, M., Rychtář, J.: A game theoretical model of kleptoparasitism with incomplete information. J. Math. Biol. **59**(5), 631–649 (2009)
6. Broom, M., Rychtář, J.: Kleptoparasitic melees - modelling food stealing featuring contests with multiple individuals. Bull. Math. Biol. **73**(3), 683–699 (2011)
7. Broom, M., Rychtář, J.: Game-Theoretical Models in Biology. CRC, Boca Raton (2013)
8. Caraco, T., Giraldeau, L.A.: Social foraging: producing and scrounging in a stochastic environment. J. Theor. Biol. **153**(4), 559–583 (1991)
9. Dubois, F., Giraldeau, L.A.: Fighting for resources: the economics of defense and appropriation. Ecology **86**(1), 3–11 (2005)
10. Dubois, F., Giraldeau, L.A., Grant, J.W.A.: Resource defense in a group-foraging context. Behav. Ecol. **14**(1), 2 (2003)
11. Enquist, M., Leimar, O.: Evolution of fighting behaviour: the effect of variation in resource value. J. Theor. Biol. **127**(2), 187–205 (1987)
12. Grimm, M.P., Klinge, M.: Pike and some aspects of its dependence on vegetation. In: Craig, J.F. (ed.) Pike: Biology and Exploitation, pp. 125–156. Chapman & Hall, London (1996)

13. Iyengar, E.V.: Kleptoparasitic interactions throughout the animal kingdom and a re-evaluation, based on participant mobility, of the conditions promoting the evolution of kleptoparasitism. Biol. J. Linn. Soc. **93**(4), 745–762 (2008)
14. Jeanne, R.L.: Social biology of the neotropical wasp mischocyttarus drewseni. Bull. Mus. Comp. Zool. **144**, 63–150 (1972)
15. Kruuk, H.: The Spotted Hyena: A Study of Predation and Social Behavior. University of Chicago Press, Chicago (1972)
16. Spear, L.B., Howell, S.N.G., Oedekoven, C.S., Legay, D., Bried, J.: Kleptoparasitism by brown skuas on albatrosses and giant-petrels in the Indian ocean. Auk **116**(2), 545–548 (1999)
17. Steele, W.K., Hockey, P.A.R.: Factors influencing rate and success of intraspecific kleptoparasitism among kelp gulls (Larus dominicanus). Auk **112**(4), 847–859 (1995)
18. Triplet, P., Stillman, R.A., Goss-Custard, J.D.: Prey abundance and the strength of interference in a foraging shorebird. J. Anim. Ecol. **68**(2), 254–265 (1999)
19. Vickery, W.L., Giraldeau, L.A., Templeton, J.J., Kramer, D.L., Chapman, C.A.: Producers, scroungers and group foraging. Am. Nat. **137**(6), 847–863 (1991)

# Chapter 14
# A Field Test of Optional Unrelated Question Randomized Response Models: Estimates of Risky Sexual Behaviors

Tracy Spears Gill, Anna Tuck, Sat Gupta, Mary Crowe, and Jennifer Figueroa

## 14.1 Introduction

Subjects tend to provide a more socially desirable response when asked about illegal or highly stigmatized behaviors [21]. This is known as social desirability response bias and can make estimating the prevalence of these behaviors problematic. One technique used to reduce this response bias is the randomized response technique (RRT). This method, introduced by Warner [22], increases subject anonymity by asking the sensitive question in an indirect manner. With a greater sense of anonymity, subjects are more likely to provide a truthful response [19]. RRT models have been used successfully to obtain accurate estimates of a variety of behaviors susceptible to response bias in self-report surveys including AIDS [2], lying and cheating [12], drug use by athletes [20], and veterinary diseases [4].

T.S. Gill (✉)
School of Nursing, University of North Carolina at Greensboro, PO Box 26170,
Greensboro, NC 27170, USA
e-mail: tgspears@unc.edu

A. Tuck • S. Gupta
Department of Mathematics and Statistics, University of North Carolina at Greensboro,
317 College Avenue, Greensboro, NC 27412, USA
e-mail: avmikh@uw.edu; sngupta@uncg.edu

M. Crowe
Department of Experiential Education, Florida Southern College, 111 Lake Hollingsworth,
Dr. Lakeland, FL 33801, USA
e-mail: mcrowe@flsouthern.edu

J. Figueroa
Department of Biology, University of North Carolina at Greensboro, 321 McIver Street,
Greensboro, NC 27402, USA
e-mail: jmfiguer@uncg.edu

J. Rychtář et al. (eds.), *Topics from the 8th Annual UNCG Regional Mathematics
and Statistics Conference*, Springer Proceedings in Mathematics & Statistics 64,
DOI 10.1007/978-1-4614-9332-7_14, © Springer Science+Business Media New York 2013

The unrelated question RRT model, proposed by Greenberg et al. [8], is a variation of Warner's original model that has been shown to be a more efficient alternative [7, 15]. In unrelated question RRT, a predetermined proportion of subjects is asked an innocuous, unrelated question with known (or unknown) mean. A randomization device is used anonymously by subjects to determine which question (sensitive or innocuous) must be answered. Since the researcher knows only the reported response, not which question was answered, the subjects anonymity in regard to the sensitive behavior is preserved. As the researcher determines proportion of subjects who receive the sensitive question, and knows the true mean of the unrelated question, the mean of the sensitive behavior can be estimated at the aggregate level.

Recently a variation of the RRT model, known as Optional RRT model, has been proposed by Gupta et al. [9]. Optional models take into account the fact that a question may be sensitive to one subject, but not sensitive to another. Subjects finding the question not personally sensitive are instructed to ignore the innocuous question, if obtained from the randomization device, and answer the sensitive research question instead. Optional models allow estimation of two parameters. In addition to population mean or prevalence, estimated by all RRT methods, optional models also estimate the sensitivity level of the underlying sensitive behavior. sensitivity level is defined as the proportion of subjects who find the question sensitive, and hence want the extra anonymity of the randomization device in answering. Knowledge of the sensitivity level is important because it allows researchers to assign better-trained interviewers for more sensitive questions. Sensitivity estimation also plays a critical role in Multi-Stage RRT models [10, 14].

Optional RRT models have shown promise in theoretical papers and computer simulations, but their performance has not been evaluated through field surveys involving real sensitive topics. This paper presents a field test of the optional unrelated-question RRT models introduced in Gupta et al. [11], covering both binary and quantitative response situations. The estimates of population mean and prevalence obtained by this method are compared to results obtained by using direct face-to-face interview method and anonymous check-box survey method. Estimates of sensitivity level are only obtained by optional RRT methods, and so cannot be compared directly to other survey methods. Our expectation is that estimates obtained from optional RRT models will match well with those of check-box survey method (assumed to represent the true status), since both provide subjects anonymity, and that results based on face-to-face interview surveys will be low. Additionally, optional RRT models will provide an estimate of sensitivity level.

## 14.2  Optional Unrelated Question RRT Models

Optional unrelated question RRT model formulas for mean estimator ($\hat{\mu}_X$), prevalence estimator ($\hat{\pi}_X$), and corresponding Sensitivity estimators ($\hat{W}_1, \hat{W}_{\pi_1}$) are provided in Gupta et al. [11] and are briefly summarized below. Figure 14.1 illustrates the process of answering a question in an Optional unrelated question RRT survey.

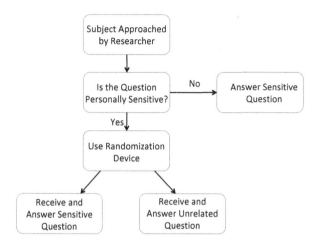

**Fig. 14.1** Flowchart depicting optional unrelated question RRT subject answering procedure

### *14.2.1  Quantitative Model*

Let $X$ be the true sensitive variable of interest with unknown mean $\mu_X$ and unknown variance $\sigma_X^2$, and $Y$ be a non-sensitive variable with known mean $\mu_Y$ and known variance $\sigma_Y^2$. Let $p$ represent probability of receiving the sensitive question from the randomization device. Let $W$ be the sensitivity level of the question. That is, a proportion $W$ of the respondents considers the question sensitive and will choose to provide a scrambled response. Others will provide a direct response with probability $(1 - W)$. This is done using color-coded cards unobserved by the respondent.

The reported response Z under this scenario is given by:

$$Z = \begin{cases} X, & \text{with probability} \quad (1 - W) + Wp \\ Y, & \text{with probability} \quad W(1 - p) \end{cases} \tag{14.1}$$

with

$$E(Z) = (1 - W)E(X) + W(pE(X) + (1 - p)E(Y)) \tag{14.2}$$

and

$$\text{Var}(Z) = [(1 - W) + Wp]E(X^2) + W(1 - p)E(Y^2) - [E(Z)]^2 \tag{14.3}$$

Using a split sample approach, the estimate of population mean $\mu_X$ is given by

$$\hat{\mu}_X = \frac{\bar{Z}_1 - \lambda \bar{Z}_2}{1 - \lambda} \quad \text{with} \tag{14.4}$$

$$\text{Var}(\hat{\mu}_X) = \frac{\text{Var}(\bar{Z}_1) + \lambda^2 \text{Var}(\bar{Z}_2)}{(1 - \lambda)^2} \tag{14.5}$$

where $\bar{Z}_i$, $(i = 1, 2)$ is the sample mean of reported responses in the $i$th sub-sample, $p_i$ is the probability of receiving the sensitive question in the $i$th sub-sample, and $\lambda = (1 - p_1)/(1 - p_2)$.

An approximation of the sensitivity estimator $(\hat{W}_1)$ is obtained by using a first order Taylor's approximation (with $A = E(\bar{Z}_1)$ and $B = E(\bar{Z}_2)$) and is given by

$$\hat{W}_1 = \frac{A - B}{\mu_Y(p_2 - p_1) + (1 - p_2)A - (1 - p_1)B} +$$

$$+ \frac{(p_2 - p_1)(\mu_Y - B)((\bar{Z}_1) - A)}{[\mu_Y(p_2 - p_1) + (1 - p_2)A - (1 - p_1)B]^2} +$$

$$+ \frac{(p_2 - p_1)(A - \mu_Y)((\bar{Z}_2) - B)}{[\mu_Y(p_2 - p_1) + (1 - p_2)A - (1 - p_1)B]^2} \qquad (14.6)$$

with

$$\mathrm{Var}(\hat{W}_1) = \left( \frac{(p_2 - p_1)(\mu_Y - B)}{[\mu_Y(p_2 - p_1) + (1 - p_2)A - (1 - p_1)B]^2} \right)^2 \frac{\sigma_1^2}{n_1} +$$

$$+ \left( \frac{(p_2 - p_1)(\mu_Y - B)}{[\mu_Y(p_2 - p_1) + (1 - p_2)A - (1 - p_1)B]^2} \right)^2 \frac{\sigma_2^2}{n_2} \qquad (14.7)$$

where

$$\sigma_i^2 = [1 - W + Wp_i]E(X^2) + W(1 - p_i)E(Y^2) - [E(Z_i)]^2, \quad i = 1, 2 \quad (14.8)$$

### 14.2.2 Binary Model

As established in Gupta et al. [11], the estimators of the prevalence of the sensitive characteristic $(\hat{\pi}_X)$ and the corresponding sensitivity level $(\hat{W}_\pi)$ are given by

$$\hat{\pi}_X = \frac{\hat{P}_{Y_1} - \lambda \hat{P}_{Y_2}}{1 - \lambda} \qquad (14.9)$$

with

$$\mathrm{Var}(\hat{\pi}_X) = \frac{\mathrm{Var}(\hat{P}_{Y_1}) + \lambda^2 \mathrm{Var}(\hat{P}_{Y_2})}{(1 - \lambda)^2} \qquad (14.10)$$

and

$$\hat{W}_\pi = \frac{\hat{P}_{Y_1} - \lambda \hat{P}_{Y_2}}{(p_2 - p_1)(\pi_Y - \hat{\pi}_X)} \qquad (14.11)$$

The expression for $\hat{W}_\pi$ in (14.11) is a ratio of two random variables. Its approximation based on the first order Taylors approximation $(\hat{W}_{\pi_1})$ and its variance is given by

$$\hat{W}_{\pi_1} = \frac{P_{Y_1} - P_{Y_2}}{\pi_Y(p_2 - p_1) + (1 - p_2)P_{Y_1} - (1 - p_1)P_{Y_2}} +$$

$$+ \frac{(p_2 - p_1)(\pi_Y - P_{Y_2})(\hat{P}_{Y_1} - P_{Y_1})}{[\pi_Y(p_2 - p_1) + (1 - p_2)P_{Y_1} - (1 - p_1)P_{Y_2}]^2} +$$

$$+ \frac{(p_2 - p_1)(P_{Y_1} - \pi_Y)(\hat{P}_{Y_2} - P_{Y_2})}{[\pi_Y(p_2 - p_1) + (1 - p_2)P_{Y_1} - (1 - p_1)P_{Y_2}]^2} \qquad (14.12)$$

with

$$\mathrm{Var}(\hat{W}_{\pi_1}) = \left( \frac{(p_2 - p_1)(\pi_Y - P_{Y_2})}{\mu_y(p_2 - p_1) + (1 + p_2)P_{Y_1} - (1 - p_1)P_{Y_2}} \right)^2 \frac{\sigma_{\pi_1}^2}{n_1} +$$

$$+ \left( \frac{(p_2 - p_1)(\pi_Y - P_{Y_2})}{\mu_y(p_2 - p_1) + (1 + p_2)P_{Y_1} - (1 - p_1)P_{Y_2}} \right)^2 \frac{\sigma_{\pi_2}^2}{n_2} \qquad (14.13)$$

where

$$\sigma_{\pi_i}^2 = \frac{P_{Y_i}(1 - P_{Y_i})}{n_i}, \quad i = 1, 2 \qquad (14.14)$$

When response bias would result in underreporting of the sensitive behavior, it is important that the mean of the unrelated question is greater than the mean of the sensitive behavior to provide subject anonymity. Also, the unrelated question must be selected such that its mean is not close to the mean of the sensitive behavior to avoid a near zero term in the denominator of the estimator for sensitivity level.

### 14.2.3   Sample Split

The sample split is based on optimal split formulas given in Gupta et al. [11]. Although there are two different sensitive questions involved in our survey (quantitative and binary), our optimal split is based on the quantitative model. The variance of $\hat{\mu}_X$ is minimized by this optimal split.

## 14.3   Previous Study

Previously, in 2011–2012, a field study was attempted by the authors using the binary version of the Optional Unrelated-Question RRT, Eqs. (14.9) and (14.11). The sensitive question used in that study was "Have you used Ritalin, Adderall, or

**Table 14.1** Stimulant
medication misuse survey
results

| Method | $\hat{\pi}_X$ | 95% CI[a] |
|---|---|---|
| Optional RRT | 0.0255 | (0.0066, 0.0443) |
| Check-box method | 0.1063 | (0.0503, 0.1413) |
| Face-to-face interview | 0.0958 | (0.0550, 0.1576) |

[a]Based on Bonferroni correction

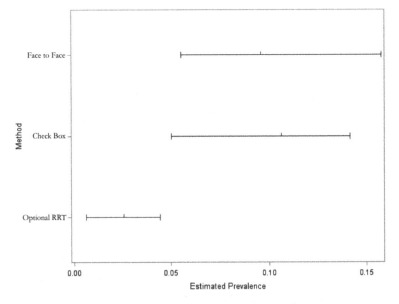

**Fig. 14.2** Estimates of prevalence of stimulant medication misuse ($\hat{\pi}_X$) by three methods, with 95 % CI based on Bonferroni correction

any other stimulant medication in the past 12 months in ways that are not prescribed by a physician?" and the innocuous unrelated question was "Were you born in the month of April?" This question was selected because the prevalence of an April birthday ($\pi_Y = 0.0822$) is similar to the prevalence of simulant medication misuse reported in other studies [6, 18]. In that study, check-box survey and direct face-to-face interview methods were used in addition to the Optional Unrelated-Question RRT method. Survey results are provided in Table 14.1 and Fig. 14.2, with confidence intervals that are based on Bonferroni correction.

Note that the survey results are counter-intuitive. Questioning subjects about sensitive behavior in a face-to-face interview setting provides the least anonymity, so this method was expected to produce the lowest estimate of prevalence. However, the face-to-face interview results are almost in line with the check-box method and the optional RRT results show much lower prevalence rate. Based on this first attempt at implementing an Optional RRT model with real subjects, several critical areas for improvement were discovered. The most important was that of selection of the unrelated, innocuous question. The prevalence of April birthdays ($\pi_Y = 0.0822$) is close to the prevalence of stimulant medication misuse reported in previous studies and estimated in this study by the two other survey methods. Additionally, surveys (described below) indicate that the population in this study does not consider this a sensitive topic. A combination of these problems likely resulted in a calculated value of $\hat{W}_\pi = 1.2284$. Additionally, this study did not incorporate an optimal split between subsamples, which will increase the variance of $\hat{\pi}_X$. All of these issues were addressed in the present study, as described below.

## 14.4   Current Study

### 14.4.1   Participants

The study was conducted at the campus of University of North Carolina at Greensboro (UNCG), a public university in the southeastern United States. Enrollment at UNCG is approximately 18,000 students [1], with 67% of undergraduates being females [16]. Eight hundred and seventy-eight subjects were recruited from undergraduate level class sections in mathematics and statistics with at least 50 enrolled students. Participation was voluntary and took place during regular class time. No incentives were given for participation. Subjects had a median age of 19 and a mean age of 20.56 (875 reporting), 66.6% were female and 33.4% male (841 reporting), 5.5% reported being married and 95.5% reported not being married (871 reporting). Distribution among class levels was: 37.2% freshman, 30.5% sophomore, 19.5% junior, 12.2% senior, and 0.5% other (876 reporting).

### 14.4.2   Question Selection

To assist in selection of topics sensitive to our participant population, a short survey was given to 55 students. Twelve sample questions were provided in the survey. Students were asked to rate how sensitive they found each question on a ten-point Likert scale. Items rated most sensitive included topics related to sexual behavior. Items rated less sensitive included topics related to alcohol and stimulant medication misuse. In selecting questions for the field test, these ratings were considered in addition to the availability of previous research on the topic so that some comparison with existing studies can be made.

One binary and one quantitative question were selected for the study. The question "Have you ever been told by a healthcare professional that you have a sexually transmitted disease?" was selected as the binary question. Previous studies report prevalence of STD among college students at 10–25 % [5,13]. The innocuous, unrelated question paired with this binary question was "Were you born between January 1st and October 31st?" which has a prevalence of approximately 83 %. The question "How many sexual partners have you had in the last 12 months?" was used as the quantitative question. Previous studies report a mean of 1.8–2.2 [3, 17] in similar populations. In this case, the innocuous unrelated question is "What is the number listed on this card?" The numbers listed on the cards ranged from 0 to 9, with a mean of 4.04.

### 14.4.3  Procedure

Enrolled classes were surveyed by one of the three methods: the RRT method described above, direct face-to-face interviewing, or anonymous check-box survey. Prior to participation, all students received information about the risks and benefits of participation in the study, the questions to be asked in the study, as well as a short lecture about RRT. Those willing to participate then completed a consent form and a survey of demographic information (age, sex, marital status, year in school). Participation was voluntary. In all the three methods, participants were informed of the sensitive questions prior to completing the consent forms. The study was overseen by UNCGs Institutional Review Board.

In class sections selected for the check-box survey method, the sensitive questions were included on the demographic information sheet. After completion of this form, participants placed the survey in a collection box. In class sections selected for the face-to-face interview method, participants approached an interviewer after completing the demographic information sheet. The participant was then directly asked each of the sensitive questions, and the response was recorded by the researcher on the demographic sheet.

In sections selected for RRT, participants were instructed to consider whether either of the questions was personally sensitive (if they would hesitate to answer the question if asked directly). Upon approaching the interviewer, participants were instructed to select a card from a deck corresponding to the binary question and one from a deck corresponding to the quantitative question. If the participant earlier determined that the question is personally sensitive, he/she was to answer the question drawn from the deck. If the participant had earlier determined that the question is not personally sensitive, that question should be answered, regardless of which question (sensitive or innocuous) was drawn from the deck. Responses were recorded by the researcher on the demographic sheet.

Participants in the RRT group were split into two subsamples according to the optimal sample split formula given in Gupta et al. [11]. Rough estimates of number of sexual partners and STD prevalence, needed for sample size determination, were obtained from previous studies, and set at $\mu_X = 2$ and $\pi_X = 0.15$. No previous studies have estimated $W_1$ or $W_{\pi_1}$, but investigation of a range of possible values for $W_1$ and $W_{\pi_1}$ revealed only slight changes in the optimal split proportions (see Tables 14.2 and 14.3). Four hundred and sixty-six participants were recruited into the RRT sample, with 354 in sub-sample 1 and 112 in sub-sample 2, giving $n_1/n = 0.7597$. This is close to the optimal splits in Tables 14.2 and 14.3 based on a total sample of 500, regardless of the sensitivity level.

**Table 14.2** Optimal split using $\mu_X = 2$

| W | $n_1$ | $n_2$ | Optimal value of $n_1/n$ |
|---|---|---|---|
| 0 | 400 | 100 | 0.8 |
| 0.25 | 388 | 112 | 0.7758 |
| 0.5 | 383 | 117 | 0.7668 |
| 0.75 | 383 | 117 | 0.7663 |
| 1 | 386 | 114 | 0.7718 |

**Table 14.3** Optimal split using $\pi_X = 0.15$

| W | $n_1$ | $n_2$ | Optimal value of $n_1/n$ |
|---|---|---|---|
| 0 | 400 | 100 | 0.8 |
| 0.25 | 387 | 113 | 0.7742 |
| 0.5 | 385 | 115 | 0.7698 |
| 0.75 | 389 | 111 | 0.7778 |
| 1 | 399 | 101 | 0.7973 |

**Table 14.4** Estimates of the mean number of sexual partners in the last 12 months

| Method | $\hat{\mu}_X$ | Sample std. dev. | 95% CI[a] | n |
|---|---|---|---|---|
| Optional RRT | 1.717 | 3.9912 | $(1.2744, 2.1596)$ | 466 |
| Check-box method | 1.680 | 2.5613 | $(1.2647, 2.0953)$ | 218 |
| Face-to-face interview | 1.130 | 1.1511 | $(0.9311, 1.3289)$ | 192 |

[a]Based on Bonferroni correction

**Table 14.5** Estimates of the STD diagnosis prevalence

| Method | $\hat{\pi}_X$ | Sample std. dev. | 95% CI[a] | n |
|---|---|---|---|---|
| Optional RRT | 0.0367 | 0.1180 | $(0.0159, 0.0576)$ | 466 |
| Check-box method | 0.0900 | 0.2862 | $(0.0438, 0.1362)$ | 220 |
| Face-to-face interview | 0.0200 | 0.1400 | $(-0.0042, 0.0442)$ | 192 |

[a]Based on Bonferroni correction

Note that the total sample size allocated to RRT group is about double of what it was for the other two groups. This was because two parameters are estimated in the optional RRT case (mean and sensitivity level), compared to a single parameter in the other cases (mean).

## 14.5   Results

Results for $\hat{\mu}_X$ and $\hat{\pi}_X$ by the three survey methods are provided in Table 14.4 and Fig. 14.3, and Table 14.5 and Fig. 14.4, respectively. Confidence intervals are based on Bonferroni correction. Results for sensitivity level ($\hat{W}$, $\hat{W}_\pi$) by optional unrelated-question RRT are listed in Table 14.6.

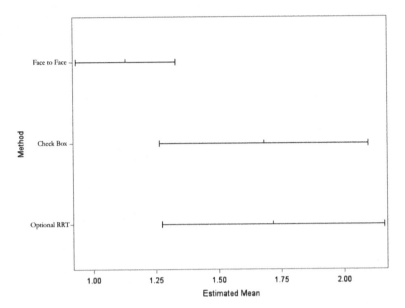

**Fig. 14.3** Estimates of number of sexual partners in the previous 12 months ($\hat{\mu}_X$) by three methods, with 95 % CI based on Bonferroni correction

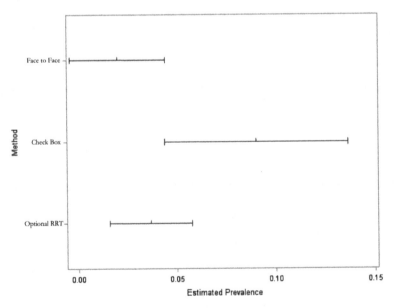

**Fig. 14.4** Estimates of STD diagnosis prevalence ($\hat{\pi}_X$) by three methods, with 95 % CI based on Bonferroni correction

**Table 14.6** Estimates sensitivity level

| Question | Sensitivity level | Est. std. dev. | 95% CI |
|---|---|---|---|
| Number of sexual partners | $\hat{W} = 0.6098$ | 0.1290 | $(0.5981, 0.6215)$ |
| STD history | $\hat{W}_\pi = 0.7730$ | 0.0196 | $(0.7712, 0.7748)$ |

## 14.6   Discussion

The estimate of $\mu_X$ obtained by Optional Unrelated-Question RRT $[\hat{\mu}_X = 1.717,$ 95 % CI $(1.2744, 2.1596)]$ is very similar to the estimate given by check-box survey method $[\hat{\mu}_X = 1.680,$ 95 % CI $(1.2647, 2.0953)]$. The lowest point estimate of $\mu_X$ is obtained by face-to-face interview $(\hat{\mu}_X = 1.130)$, which is expected since this method provides the least anonymity. However, the upper bound of the 95 % CI for this group $(0.9311, 1.3289)$ does overlap with those of the other two methods, but only slightly. For the binary method, the estimate of $\pi_X$ is highest when obtained by check-box survey $[\hat{\pi}_X = 0.0900,$ 95 % CI $(0.0900, 0.1362)]$, and lowest in face-to-face interviews $[\hat{\pi}_X = 0.0200,$ 95 % CI $(-0.0042, 0.0442)]$ with Optional Unrelated RRT being in the middle, as expected $[\hat{\pi}_X = 0.0367,$ 95 % CI $(0.0159, 0.0576)]$, but there is overlap in the 95 % confidence intervals of all the three methods. Perhaps a larger sample would be needed to see better delineation between the three methods. The results for number of sexual partners and STD prevalence obtained by optional unrelated question RRT and check-box survey are in line with estimates obtained by previous studies [3, 5, 13, 17].

As sensitivity level is estimated only by Optional RRT methods, and this is the first field test of an Optional RRT model, there is no way to make a direct comparison of these results. The sensitivity level of the question on number of sexual partners $(\hat{W} = 0.6098)$ and STD history $(\hat{W}_\pi = 0.7730)$ is consistent with the fact that these questions are generally sensitive and one would not feel comfortable answering them directly. Also the results of this field test study are consistent with the mathematical and computer simulation results presented in Gupta et al. [11].

**Acknowledgments** This work was supported by NSF grants DMS 0850465 and DBI 0926288.

## References

1. About UNCG: In: Discover UNCG. University of North Carolina at Greensboro. http:// admissions.uncg.edu/discover-about.php (2013). Cited 15 Feb 2013
2. Arnab, R., Singh, S.: Randomized response techniques: an application to the Botswana AIDS impact survey. J. Stat. Plan. Infer. **140**, 941–953 (2010)
3. Baldwin, J.D., Whitely, S., Baldwin, J.I.: The effect of ethnic group on sexual activities related to contraception and STDs. J. Sex. Res. **29**, 189–205 (1992)

4. Cross, P., Edwards-Jones, G., Omed, H., Williams, A.P.: Use of a randomized response technique to obtain sensitive information on animal disease prevalence. Prev. Vet. Med. **96**, 252–262 (2010)

5. Department of Health and Human Services: Surveillance summaries. Morb. Mortal. Wkly. Rep. (Centers for Disease Control and Prevention) **58**, 1–29 (2009). http://www.cdc.gov/mmwr. Cited15Dec2012

6. DeSantis, A.D., Webb, E.M., Noar, S.M.: Illicit use of prescription ADHD medications on a college campus: a multimethodological approach. J. Am. Coll. Health **57**, 315–323 (2008)

7. Dowling, T.A., Shachtman, R.: On the relative efficiency of randomized response models. In: Institute of Statistics Mimeo Series, vol. 811. University of North Carolina, Greensboro (1972)

8. Greenberg, B.G., Abul-Ela, A.-L.A., Simmons, W.R., Horwitz, D.G.: The unrelated question randomized response model: theoretical framework. J. Am. Stat. Assoc. **64**, 520–539 (1969)

9. Gupta, S.N., Gupta, B., Singh, S.: Estimation of the sensitivity level of personal interview survey questions. J. Stat. Plan. Infer. **100**, 239–247 (2002)

10. Gupta, S.N., Mehta, S., Shabbir, J., Dass, B.K.: Some optimality issues in estimating two-stage optional randomized response models. Am. J. Math.-S. **31**, 1–12 (2011)

11. Gupta, S.N., Tuck, A., Gill, T.S., Crowe, M.: Optional unrelated-question randomized response models. Involve J. Math. **6**, 483–492 (2013)

12. Lensvelt-Mulders, G.J.L.M., Boeiji, H.R.: Evaluation of compliance with a computer assisted randomized response technique: a qualitative study into the origins of lying and cheating. Comput. Hum. Behav. **23**, 591–608 (2007)

13. MacDonald, N.E., Wells, G.A., Fisher, W.A., Warren, W.K., King, M.A., Doherty, J.A., Bowie, W.R.: High-risk STD/HIV behavior among college students. J. Am. Med. Assoc. **263**, 3155–3159 (1990)

14. Mehta, S., Dass, B.K., Shabbir, J., Gupta, S.N.: A three-stage optional randomized response model. J. Stat. Theor. Pract. **6**, 417–427 (2012)

15. Moors, J.J.A.: Optimization of the unrelated question randomized response model. J. Am. Stat. Assoc. **66**, 627–629 (1971)

16. Office of Equity, Diversity, and Inclusion: Diversity at UNCG. University of North Carolina at Greensboro. http://oedi.uncg.edu/diversity (2013). Cited 15 Feb 2013

17. Reinisch, J.M., Hill, C.A., Sanders, S.A., Ziemba-Davis, M.: High-risk sexual behavior at a Midwestern university: a confirmatory study. Fam. Plann. Perspect. **27**, 79–82 (1995)

18. Shillington, A.M., Reed, M., Lange, J., Clapp, J., Henry, S.: College undergraduate Ritalin abusers in southwestern California: protective and risk factors. J. Drug Issues **36**, 999–104 (2006)

19. Stem, D.E., Bozman, C.S.: Respondent anxiety reduction with the randomized response technique. Adv. Consum. Res. **15**, 595–599 (1988)

20. Striegal, H., Ulrich, R., Simon, P.: Randomized response estimates for doping and illicit drug use in athletes. Drug Alcohol Depend. **106**, 230–232 (2010)

21. Tourangeau, R., Yan, T.: Sensitive questions in surveys. Psychol. Bull. **133**, 859–833 (2007)

22. Warner, S.L.: Randomized response: a survey technique for eliminating evasive answer bias. J. Am. Stat. Assoc. **60**, 63–69 (1965)

# Chapter 15
# A Spatially Organized Population Model to Study the Evolution of Cooperation in Species with Discrete Life-History Stages

Caitlin Ross, Olav Rueppell, and Jan Rychtář

## 15.1 Introduction

The evolution of cooperation and altruism has intrigued scientists from more than a century because it superficially seems to have individuals act against the paradigm of Darwinian fitness maximization [5]. Nevertheless, cooperative and altruistic behavior occurs in a number of different taxa [8]. In most species, cooperation and altruism are linked to kin selection [10], the argument that individuals can gain fitness by helping related individuals reproduce [11, 12]. Criticism of kin selection as the only underlying concept of inclusive fitness theory has led to the insight that spatial structures of natural populations are key to the evolution of cooperation and altruism [16].

The evolution of cooperation and altruism has been addressed by a simple game called the Prisoner's Dilemma [19, 25]. The game is one of the most widely studied games in biology [15] and it is used in different variations of increasing complexity [1, 2, 7]. In its simplest case, the Prisoner's Dilemma game involves two individuals that interact once and can either cooperate with each other or try to deceive the other individual. The evolutionary benefit from their interaction is

C. Ross
Department of Computer Sciences, The University of North Carolina
at Greensboro, Greensboro, NC 27402, USA
e-mail: cjkintz@uncg.edu

O. Rueppell (✉)
Department of Biology, The University of North Carolina
at Greensboro, Greensboro, NC 27402, USA
e-mail: olav_rueppell@uncg.edu

J. Rychtář
Department of Mathematics and Statistics, The University of North Carolina
at Greensboro, Greensboro, NC 27402, USA
e-mail: rychtar@uncg.edu

J. Rychtář et al. (eds.), *Topics from the 8th Annual UNCG Regional Mathematics and Statistics Conference*, Springer Proceedings in Mathematics & Statistics 64, DOI 10.1007/978-1-4614-9332-7_15, © Springer Science+Business Media New York 2013

determined according to a payoff matrix, where the cooperator receives less benefit than the defector, yet the benefit for mutual cooperation is higher than the benefit for mutual defection. Despite defectors outcompeting cooperators in this simplest scenario, cooperative strategies can evolve in a number of variations of the Prisoners Dilemma, including games played in a structured population [6, 17, 18].

Population structure is typically introduced as a square lattice with interaction between neighboring nodes [22]. Under most circumstances, the structuring facilitates the evolution of cooperation because cooperators interact more often with cooperators than defectors. An important issue has been the role of diversity among players. Variation in competitive ability may or may not increase the probability for cooperation [14], while cooperation is promoted by variation in social variables [21], reproductive ability [23], and reproductive timing [27]. Equally relevant is the incorporation of player aging because all biological species age at some rate, which correlates with a number of variables, including social status and mortality rates [9]. Specifically, aging has been incorporated mostly as a maturation process: With the age of players the payoff [26] or the strategy transfer ability [24] may increase, which facilitates the evolution of cooperation in both cases. However, aging is also accompanied by many detrimental changes, including a loss of function and increasing mortality risk [20].

Evolutionarily most relevant across all species are age-dependent changes in reproductive status. In all biological species, individuals have to grow and mature until they reach reproductive age but the timing of the onset of reproduction is highly variable. Biological species also differ in the duration of their reproductive phase: While some species only reproduce once in their life, the reproductive phase represents the largest part of the lifespan in other species. A few species, including humans, have even evolved a post-reproductive lifespan that can have a considerable duration after the last reproductive event. These fundamental differences in life history structure correlate to some extent with the social organization of the species. Typically, social species have a later onset of reproduction and the phenomenon of a post-reproductive lifespan is only known from social species. As sociality is dependent on cooperation and altruism, these observations pose the question whether life history is a consequence of social organization or whether life history structure could also influence the evolution of cooperation and altruism. While several studies have addressed the first possibility [3, 4], the alternative has not yet been addressed.

Therefore, we develop a model of a spatially organized population of individuals in a square lattice that interact with their neighbors in either cooperative or non-cooperative way. These individuals transition from a potentially pre-reproductive to a reproductive and then potentially to a post-reproductive stage in a semi-deterministic fashion and they also die with a certain probability. Death results in a reproductive opportunity of all surrounding individuals and the individual with the highest payoff from all combined local Prisoner's Dilemma games will reproduce an offspring of identical phenotype into the empty spot. Our model confirms that cooperation can increase from a small cluster of cooperators in this spatially structured population, dependent on the relative cost of cooperation. In addition, we find the likelihood that cooperation is fixed in the population also to be dependent on the population structure and on the life histories.

## 15.2  Methods

To simulate the spatial structure of the population, we assume the individuals live on a regular $L \times L$ square lattice with periodic boundaries. We used $L = 128$ in our simulations as initial tests showed that such $L$ is not only large enough to avoid small populations effects but also small enough for the simulations to finish in a reasonable time. We consider two types of neighborhoods where each individual has (a) $n = 4$ neighbors (north, east, south, west), or (b) $n = 8$ neighbors (north, northeast, east, southeast, south, southwest, west, northwest), sometimes called Moore neighborhood [13].

Each node of the square lattice is either empty, or occupied by an individual (in some stage of its life). The (average) durations (measured in reproductive seasons) of those stages, denoted by $d_{pre}, d_{rep}, d_{post}$, are parameters of the simulations. By $d_{life} = d_{pre} + d_{rep} + d_{post}$ we denote the (average) life span. We refer to every fixed combination of the durations as an *age setup*. We used two different age setups: $(0, 3, 0)$ and $(1, 1, 1)$, representing individuals having only a reproductive stage and individuals having all three stages with equal durations, respectively. The aging of individuals is assumed to be biological (stochastic) and will be explained in detail below.

For each fixed age setup, the population is initialized so that the frequencies of individuals in appropriate life stages are close to the equilibrium state where the number of individuals in any particular stage is proportional to the duration of the stage. Specifically, every node of the $L \times L$ lattice is set empty with probability $(1 + d_{life})^{-1}$, or occupied by an individual in the stage $x$ (pre-reproductive, reproductive, or post-reproductive) with probability $d_x(1 + d_{life})^{-1}$. To initialize, almost all individuals are considered to be defectors; only a small number of individuals living in a small $8 \times 8$ lattice are set as cooperators. This means that roughly only $8^2/128^2 \approx 0.004$ fraction of individuals are cooperators. We have adopted this cluster seeding because when we initially tested random assignments of the strategies, the clusters have formed fast (it follows from the updating rules described below that clusters form naturally as individuals do not move and offspring are always placed next to the parent).

We then use Monte Carlo simulation to update the population in a series of elementary steps. It will follow that when we do $L^2$ such steps, every individual on average ages by one reproductive season and the average life of an individual is thus $d_{life}L^2$ of such elementary updates. We will refer to so many updates as a *generation*. For the purpose of this paper, we have run the simulations until one strategy dominated and the other vanished; or until we have updated for a total of

5,000 generations. We have chosen 5,000 generations after initial testing when we observed that when both strategies coexist after 5,000 generations, they still very likely coexist even after 30,000 generations.

Every elementary step starts by randomly selecting a node in the $L \times L$ square lattice. If the node is occupied by an individual in stage $x$ (pre-reproductive, reproductive or post-reproductive), then the individual "stochastically ages." It means that the individual moves to the next life stage (or "dies" and the node becomes empty if it is currently in the last stage of its life) with probability $d_x^{-1}$. This assures that the duration of a particular stage is on average $d_x$ updates of the node. Since every node is updated on average once in $L^2$ updates, $d_x$ updates of the particular node roughly corresponds to aging by $d_x$ reproductive seasons. We note, however, that for the age setup $(0, 3, 0)$, the exact length of any life stage is stochastic and not deterministic and the aging is thus biological.

If the randomly selected node is not occupied, then one of the following things will happen. If there is no individual in a reproductive stage in the neighborhood of the focal (selected) node, then the focal node will remain empty. Otherwise, one of the reproductive individuals in the neighborhood is selected to be a parent and it then places an offspring into the focal node. The offspring will inherit the parent's strategy (i.e., it will be cooperator if and only if the parent is) and it starts at the first life stage (pre-reproductive if $d_{pre} > 0$, or reproductive if $d_{pre} = 0$).

The parent selection is done at random and proportional to the fitness which is calculated as follows. If the prospective parent has $N_C$ cooperators (and $N_D$ defectors and $N_E$ empty spaces) in the neighborhood, then its fitness is given by

$$f = 1 + \begin{cases} N_C, & \text{if it is cooperator} \\ bN_C, & \text{if it is defector.} \end{cases} \tag{15.1}$$

where 1 corresponds to a background fitness. A positive background fitness has to be included as otherwise individuals with no cooperative neighbor (i.e., typically the defectors) would not reproduce. We have tested for several values of the background fitness, but we did not spot any significant differences, so we settled for background fitness of 1. The above formula (15.1) means that the fitness is calculated based on the evolutionary prisoner's dilemma game that the prospective parents play with all of *their* neighbors. The payoff matrix of the game is

$$\begin{array}{cc} & \begin{array}{cc} \text{Cooperate} & \text{Defect} \end{array} \\ \begin{array}{c} \text{Cooperate} \\ \text{Defect} \end{array} & \begin{pmatrix} 1 & 0 \\ b & 0 \end{pmatrix} \end{array} \tag{15.2}$$

and has already been studied and used in a similar context in [24]. The advantage of such a simple matrix is that $b$ is the only parameter; and for $1 < b \leq 2$ all the important aspects of prisoner's dilemma game are preserved.

For every age setup, we have run 100 simulations described above. If one of the strategies reached a fixation, we noted the time of that event (in generations).

For $n = 8$, the age setup $(0, 3, 0)$ and $b$ between 1.3 and 1.7, both cooperators and defectors still existed after 5,000 generations, we noted the time as 5,000. The coexistence actually persisted for quite a large number of generations and we thus calculated the probability of persistence, which we calculated as the fraction of time the cooperators made at least 5% population at the end of the generation 5,000.

## 15.3   Results

For $n = 4$, the fixation probability as well as the average time to fixate did not depend significantly on $b$, see Figs. 15.1 and 15.2. In fact, for the age setup $(0, 3, 0)$, the cooperators always fixated with probability 1 and typically in around 200–250 generations (the time increased from about 250 to 450 as $b$ increased from 1.8 to 2). For the age setup $(1, 1, 1)$, the fixation probability decreased slightly with $b$ from about 0.9 (for $b = 1$) to 0.8 (for $b = 2$) and the time to fixation was gradually increasing from about 450 generations at $b = 1$ to 550 generations at $b = 2$.

For $n = 8$, the fixation was close to 1 for small values of $b < 1.3$ and the cooperators fixate relatively fast (in less than 200 generations) . For the age setup $(1, 1, 1)$, the fixation stays close to 1 even for $b < 1.7$. For the age setup $(0, 3, 0)$, the cooperators never fixated when $b > 1.3$. However, they still sometimes persisted in the population for $b$ as high 1.7 and the persistence decreased gradually as $b$ increased from 1.3 to 1.7. The cooperators were eliminated for $b > 1.7$. For the age setup $(1, 1, 1)$, the cooperators did not fixate for $b > 1.7$, but persisted in the population with probability $1/4$ even for $b = 2$.

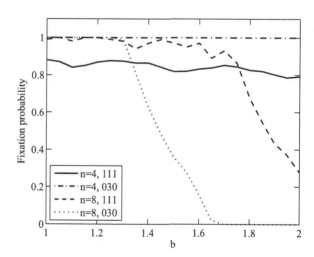

**Fig. 15.1**  Fixation probability of cooperators as it depends on $b$, the neighborhood size and the life history

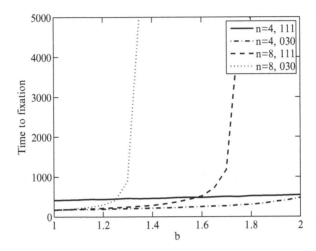

**Fig. 15.2** The time to fixation of cooperators as it depends on $b$, the neighborhood size and the life history

## 15.4   Conclusions and Discussion

Overall, our model confirms that cooperation can evolve in spatially structured populations, even when cooperation has a substantial intrinsic disadvantage compared to defection [22–24]. Our preliminary analysis of this model have shown that both the neighborhood size and the life history affect the evolution of cooperation and altruism.

A large neighborhood size decreases the probability of fixation of cooperators in the population. Cooperation still fixates in the population with a high probability (over 80 %) for small neighborhoods ($n = 4$) even for large $b$ (when the defectors have a big advantage), but does not fixate (or only with a very low probability) for larger ($n = 8$) neighborhoods. The same phenomenon in similar settings has already been observed in [13].

The results of our model that restricts the neighborhood size to four indicate that the advantage of defectors ($b$) does not play a significant role and cooperators fixate with a roughly constant probability regardless of $b$. This can be explained as follows. A defector can only obtain a higher fitness than a cooperator when at least one cooperator is in its neighborhood. However, there are only four neighboring spots. One spot is empty, another is typically still occupied by the parent (that uses same strategy as the focal individual) and thus a defector practically cannot have more than two cooperators in its neighborhood and very often has none. The spatial structure leads to clustering of the cooperators and thus a typical cooperator has two or three cooperators in its neighborhood, outperforming the defectors regardless of the value of $b$.

In accordance with other models, we observed that the life-history affects the evolution of cooperation [24, 26, 27]. However, the effect interacts with the neighborhood size. In a population that allows interaction with only four neighbors, the fixation probability of cooperators across all values of b is higher and the average time to fixation lower when the reproductive phase is long (030 model) compared to a life history with an even duration of pre-reproductive development, reproduction, and post-reproductive aging phase (111 model). In the second population structure that allows interactions with eight neighbors, the result is reversed across all values of b. The reasons for these outcomes are not quite clear and more research will have to be done. However, we note here that the stochasticity of the population behavior is reduced by both a longer reproductive phase and a larger neighborhood size, because both prevent local stochastic extinctions of the population due to a lack of reproductive active neighbors when a reproductive opportunity arises.

**Acknowledgments** The research was supported by an NSF grants DMS 0850465 and DBI 0926288. The authors wish to thank Kayla Jackson for her initial input in the development of the early stages of the model.

# References

1. Axelrod, R.: The Evolution of Cooperation. Basic Books, New York (1984)
2. Axelrod, R., Hamilton, W.D.: The evolution of cooperation. Science 27 **211**(4489) pp. 1390–1396 (1981) DOI: 10.1126/science.7466396
3. Carey, J.R.: Demographic mechanisms for the evolution of long life in social insects. Exp. Gerontol. **36**(4), 713–722 (2001)
4. Carey, J.R., Judge, D.S.: Life span extension in humans is self-reinforcing: a general theory of longevity. Popul. Dev. Rev. **27**(3), 411–436 (2001)
5. Dawkins, R.: The Selfish Gene. Number 199. Oxford University Press, Oxford (1989)
6. Doebeli, M., Hauert, C.: Spatial structure often inhibits the evolution of cooperation in the snowdrift game. Nature **428**(6983), 643–646 (2004)
7. Doebeli, M., Hauert, C.: Models of cooperation based on the prisoner's dilemma and the snowdrift game. Ecol. Lett. **8**(7), 748–766 (2005)
8. Dugatkin, L.A.: Cooperation Among Animals: An Evolutionary Perspective. Oxford University Press, New York (1997)
9. Finch, C.E.: Longevity, Senescence, and the Genome. University of Chicago Press, Chicago (1994)
10. Foster, K.R., Wenseleers, T., Ratnieks, F.L.W.: Kin selection is the key to altruism. Trends Ecol. Evol. **21**(2), 57–60 (2006)
11. Hamilton, W.D.: The genetical evolution of social behaviour. I. J. Theor. Biol. **7**(1), 1–16 (1964)
12. Hamilton, W.D.: The genetical evolution of social behaviour. II. J. Theor. Biol. **7**(1), 17–52 (1964)
13. Ifti, M., Killingback, T., Doebeli, M.: Effects of neighbourhood size and connectivity on spatial continuous prisoner's dilemma. J. Theor. Biol. **231**, 97–106 (2004). arXiv preprint q-bio/0405018
14. Mesterton-Gibbons, M., Sherratt, T.N.: Information, variance and cooperation: minimal models. Dyn. Games Appl. **1**(3), 419–439 (2011)

15. Nowak, M.A.: Evolutionary Dynamics, Exploring the Equations of Life. Belknap Press of Harvard University Press, Cambridge (2006)
16. Nowak, M.A.: Five rules for the evolution of cooperation. Science **314**(5805), 1560–1563 (2006)
17. Nowak, M.A., May, R.M.: Evolutionary games and spatial chaos. Nature **359**(6398), 826–829 (1992)
18. Ohtsuki, H., Hauert, C., Lieberman, E., Nowak, M.A.: A simple rule for the evolution of cooperation on graphs and social networks. Nature **441**(7092), 502–505 (2006)
19. Poundstone, W.: Prisoner's Dilemma: John von Neumann, Game Theory and the Puzzle of the Bomb. Anchor Books, New York (1992)
20. Rueppell, O., Christine, S., Mulcrone, C., Groves, L.: Aging without functional senescence in honey bee workers. Curr. Biol. **17**(8), R274–R275 (2007)
21. Santos, F.C., Pinheiro, F.L., Lenaerts, T., Pacheco,J.M.: The role of diversity in the evolution of cooperation. J. Theor. Biol. **299**, 88–96 (2011)
22. Szabó, G., Tőke, C.: Evolutionary prisoners dilemma game on a square lattice. Phys. Rev. E **58**(1), 69 (1998)
23. Szolnoki, A., Perc, M., Szabó, G.: Diversity of reproduction rate supports cooperation in the prisoner's dilemma game on complex networks. Eur. Phys. J. B-Condens. Matters Comp. Syst. **61**(4), 505–509 (2008)
24. Szolnoki, A., Perc, M., Szabó, G., Stark, H.-U.: Impact of aging on the evolution of cooperation in the spatial prisoners dilemma game. Phys. Rev. E **80**(2), 021901 (2009)
25. Tucker, A.W.: On jargon: the prisoner's dilemma. UMAP J. **1**(101) (1980).
26. Wang, Z., Zhu, X., Arenzon, J.J.: Cooperation and age structure in spatial games. Phys. Rev. E **85**(1), 011149 (2012)
27. Wu, Z.-X., Rong, Z., Holme, P.: Diversity of reproduction time scale promotes cooperation in spatial prisoners dilemma games. Phys. Rev. E **80**(3), 036106 (2009)

# Chapter 16
# Analysis of Datasets for Network Traffic Classification

Sweta Keshapagu and Shan Suthaharan

## 16.1 Introduction

The classification of network traffic has become an important requirement for network security solutions due to significant growth in Internet usage, with many applications that led to a large variety of traffic flowing over computer networks. Efficient classification algorithms can help manage network traffic and analyze security risk to help Internet Service Providers (ISP) provide high quality of service to their customers. The various types of traffic that can be found over the network include web traffic (http), secure web traffic (https), email traffic (imap, POP3, smtp, etc.), and file transfer traffic (ftp). Among these, http and https constitute the majority of traffic flowing through the network, where the http traffic, in particular, shows significant vulnerability. Hence these two traffic types must be captured at the gateway (e.g. firewall) of a computer network and classified for further analysis. This will also allow https traffic to pass through the firewall faster and improve quality of service requirements.

Machine Learning techniques have been extensively used for classification problems in network security applications due to their ability to learn statistical and mathematical properties of network traffic. They are categorized into unsupervised and supervised learning techniques [9]. The statistical similarity and the differences of the traffic characteristics are used by unsupervised learning techniques to isolate traffic classes. Hence it does not use training (i.e. labeled) datasets, but examines the properties within the incoming dataset and classifies that dataset itself. Hence unsupervised learning techniques are suitable when the training datasets are not available. Supervised learning techniques use a training datasets and produce

S. Keshapagu (✉) • S. Suthaharan
Department of Computer Science, The University of North Carolina
at Greensboro, Greensboro, NC 27412, USA
e-mail: s_keshap@uncg.edu; s_suthah@uncg.edu

J. Rychtář et al. (eds.), *Topics from the 8th Annual UNCG Regional Mathematics and Statistics Conference*, Springer Proceedings in Mathematics & Statistics 64, DOI 10.1007/978-1-4614-9332-7_16, © Springer Science+Business Media New York 2013

classifiers based on the statistical and mathematical properties learned. One of the supervised learning techniques that have been extensively used in network security research is the Support Vector Machine (SVM) [3]. Several SVM approaches have been proposed in Machine Learning research focusing on different classification applications [2, 6, 8, 13].

The classification accuracy of the SVM heavily depends on the feature variable selection, feature extraction, and distance metric learning adopted in the classification process. If irrelevant feature variables are selected and SVM is applied in the feature space, defined by these feature variables, then the classification will not give acceptable results. Similarly, if the extracted features do not accurately characterize the traffic types, then the classification results become inappropriate. Finally, if the distance metric does not measure the distance between the data points, then the separation of traffic classes becomes difficult. Hence, feature variable learning, feature extraction learning, and distance metric learning algorithms are required and they will help the SVM achieve high classification accuracy. This paper only deals with the feature extraction and distance metric learning.

We recently studied the LBNL [10] datasets using visualization tools (i.e. simple mathematical graphs) and noticed some interesting properties of http and https traffic. We intuitively—based on the networking knowledge—selected TCP window-size and packet-length as the two variables and plotted them on a two-dimensional space defined by these two variables. Interestingly, we noticed that plotting the two traffic datasets formed rectangular shape patterns. This geometric property motivated us to study the datasets further and explore feature extraction and distance metric learning. The geometric properties of the rectangular patterns and their class-separate properties are further investigated based on the features selected.

## 16.2 Background

The importance of traffic classification has been recently highlighted by Dainotti et al. [4] in their recent paper. They discussed in detail the issues and challenges that make traffic classification a difficult problem and suggested some strategies that may help overcome classification of Internet traffic. Based on their studies they provided six recommendations to enhance traffic classification systems, including the development of classification techniques with strong experimental and empirical studies for validation using diversified network traffic datasets. Since the LBNL datasets provide a strong ground with diversified traffic data and classification challenges, these datasets were selected for training and testing of the proposed learning approaches. While the validations of classifications are important, the techniques that are used to represent the traffic types are also equally important to support strong experimental and empirical studies.

One of the techniques that may be useful for classification of network traffic is the Machine Learning (ML) technique. ML techniques can be used to learn the characteristics of various network traffic types using labeled (training) datasets and classify traffic using the knowledge learned. The traffic characteristics can be represented by statistical properties and thus the statistical theory helps the Machine Learning research. For example, Zhang et al. [17] studied the effectiveness of nearest-neighbor technique for classification of network traffic and proposed a nonparametric approach incorporating correlation properties of traffic types to classify network traffic. Similarly, Zuev and Moore [20] studied the Bayes estimators and used a supervised Nave Bayes estimator to classify network traffic. However, the classification accuracy achieved by this method is not satisfactory, considering the requirements for current applications such as network security and network management. Another statistical approach has been proposed by Carela-Espaol et al. [1] and in their approach samples of NetFlow data were used. They adopted the C4.5 ML technique [14] and studied its performance using NetFlow data and demonstrated performance improvements with packet sampling techniques. They achieved a high accuracy with very low sampling rate. However, they stated that, during the training phase, the high accuracy cannot be achieved when a low sampling rate is used without adopting their packet sampling method.

The ML techniques generally depend on three representation learning tasks: feature selection learning, feature extraction learning, and distance metric learning [15]. Feature selection learning has been studied extensively in ML research and used for traffic classification in recent years. For example, Zhao et al. [18] studied the performance of ML techniques for classification of P2P network traffic and highlighted that the current feature selection approaches are not suitable for online traffic classification. Hence, they proposed a real-time feature selection approach and calculated the features on the fly. In another paper, Zhang [16] introduced a new metric called Weighted Symmetrical Uncertainty (WSU) and used this metric and a wrapper method [12] to select relevant features. The WSU metric was defined using a weighted entropy approach. Another feature selection approach is proposed by Zhen and Qiong [19] and they used an information theory approach to determine the bias of a feature towards a particular traffic class. Then, they proposed a feature selection method called BFS which reduces the number of features selected in order to simplify the problems associated with the multiple traffic class classification.

Another issue related to network traffic classification was reported by Suchul Lee et al. [11] in their paper. They indicated that most of the traffic classification techniques developed and presented in the public domain did not use the standard Benchmark tool for testing, and therefore, development of benchmark tools for the evaluation of classification techniques is required. They also presented a benchmark tool that can provide an objective comparison between classifiers. The Support Vector Machine can be considered as a benchmark tool for the traffic classification in the network security literature because it has been studied extensively in the ML research by developing several versions of SVM.

In this paper we used the Lagrangian SVM (LSVM) [13] as the benchmark classification algorithm. It has been shown that the SVM can be trained to get very high classification accuracy with an iterative training. However its suitability to online traffic classification is questionable due to its complex mathematical formulation and the need for good support vectors. One of the recent applications of SVM to TCP traffic classification is presented by Este et al. [5] in their paper. TCP classification is considered as one of the important requirements in the Internet because of its use in many network protocols like http, https, and ftp. Hence, research in TCP traffic classification using the popular ML techniques like SVM is relevant to the current technology requirements. Among the many types of TCP traffic, http and https are commonly used protocols. Hence, we address the classification of these two traffic types.

## 16.3  Proposed Approach

In this section, we propose an approach to classify http and https data using their geometric properties determined from the window size and packet length. As stated earlier, selecting features is the first and foremost step in classification problems. When a packet flows between two end points, it carries information such as source IP address, destination IP address, packet-length, protocol, and window-size. Among these, packet-length and window-size are readily available and is used for TCP hand-shaking mechanism. Hence, the use of this existing information can provide computational advantages, which is one of the reasons we selected these features for representation learning. Another reason is that, based on our findings from the analysis of LBNL datasets, http and https satisfy the rectangular geometric shapes, when the variables window-size and packet-length are plotted against each other. These geometric characteristics of the two traffic types are illustrated in Fig. 16.1.

We can clearly see the overlapping rectangles for these two traffic types. With this overlapping structure, the SVM-based classification algorithms will lead to very high false positives for both traffic classes. As an example, the application of LSVM to this dataset resulted in the classification shown in Fig. 16.2. Hence, representation learning algorithms should be adopted as a preprocessing mechanism before the application of SVM-based techniques. The rectangular patterns of the http and https traffic classes help us develop a representation learning model (feature extraction and distance metric) with tunable parameters and train the model for class separation.

We modeled these two traffic patterns using the algorithm described below. The model uses four parameters and they are the coordinate point $(c_1, c_2)$ that represents the center of the overlapping rectangular region, the distance $d_1$ that determines the left-and-right displacement of http traffic (for significant separation) and the distance $d_2$ that determines the up-and-down displacement of https traffic (for significant displacement). These four parameters will be learned using labeled datasets based on the following logic:

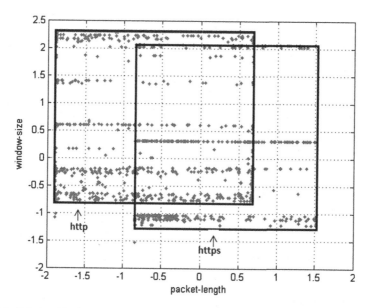

**Fig. 16.1** Relationships between the packet-length and window-size for http and https traffic

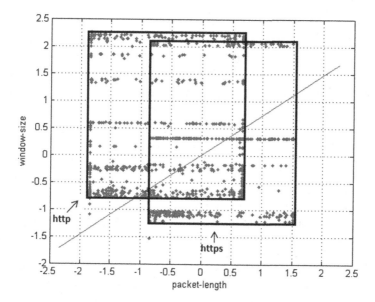

**Fig. 16.2** Application of LSVM classification to the two traffic classes

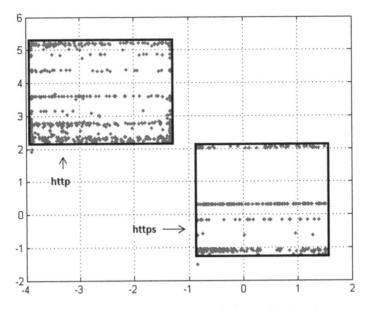

**Fig. 16.3** Initial preprocessing—isolation of the rectangles (two traffic classes)

IF (there are more data points of a particular rectangular pattern than the data points of the other rectangle on the left side of the center point $c_1$) THEN shift that rectangle left by $d_1$ ELSE shift that rectangle right by $d_1$ ENDIF.

IF (there are more data points of a particular rectangular pattern than the data points of the other rectangle on the upper side of the center point $c_2$) THEN shift that rectangle up by $d_2$ ELSE shift that rectangle down by $d_2$ ENDIF.

The goal of this model is to extract the geometric features $(c_1, c_2)$, $d_1$, $d_2$ of the rectangles from the training dataset that will help separate the rectangles of the two traffic classes, and then define a distance metric between the extracted parameters $d_1$, $d_2$ to transfer that knowledge to the classification algorithm. An example of the isolated rectangles of the http and https traffic is presented in Fig. 16.3. Then LSVM is applied to this preprocessed dataset for classification.

The features extracted from this separation of classes are $(-0.089, 0.602)$, 1.0, 1.0. We can visually see that the two classes are clearly separated with these extracted features. However, when we applied LSVM on these isolated rectangles, we get the classification shown in Fig. 16.4. The isolation of two classes causes problems to LSVM due to the inconsistent spread of data points. This classification resulted in high false positives for https traffic, but not for http traffic. Hence, to achieve high accuracy in classification, it is important to learn the model using the

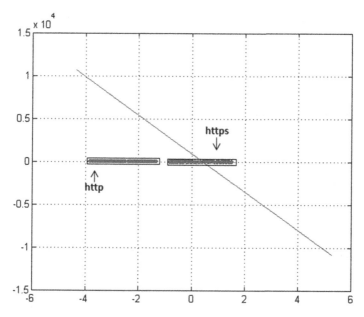

**Fig. 16.4** Applied LSVM on isolated rectangles (rescaled Y-axis for the purpose of highlighting the SVM classifier)

labeled dataset with feature extraction and distance metric learning. The continuous learning of feature extraction and distance metric at different learning phases with a cross validation technique [7] is presented in the next section.

## 16.4   Experimental Results

In this section we demonstrate the proposed feature extraction and distance metric learning models and show that the classification problem associated with SVM approaches can be eliminated by learning the geometric parameters of the rectangular patterns. To demonstrate learning and the validation of the algorithm, we divided the training dataset as 80% for learning and cross validation and 20% for testing.

### 16.4.1   Training Phase

In the training phase, we first learn the model parameters by preprocessing (i.e., the representation learning approach) the traffic data and updating the separation of the rectangles iteratively. This is achieved by choosing appropriate values for distance metric parameters $d_1$ and $d_2$ at different learning phases. This learning

**Table 16.1** Tenfold False Positives and the average for http traffic - Learning Phase I

| Tenfolds | 1 | 2 | 3 | 4 | 5 | 6 | 7 | 8 | 9 | 10 | Average |
|---|---|---|---|---|---|---|---|---|---|---|---|
| FPP | 62.76 | 73.64 | 73.09 | 70.24 | 74.67 | 71.71 | 71.82 | 71.16 | 72.93 | 69.11 | 71.11 |

**Table 16.2** Tenfold False Positives and the average for https traffic - Learning Phase I

| Tenfolds | 1 | 2 | 3 | 4 | 5 | 6 | 7 | 8 | 9 | 10 | Average |
|---|---|---|---|---|---|---|---|---|---|---|---|
| FPP | 48.86 | 57.74 | 61.46 | 57.3 | 52.08 | 52.12 | 52.31 | 50.36 | 49.39 | 51.78 | 53.44 |

**Table 16.3** Tenfold False Positives and the average for http traffic - Learning Phase II

| Tenfolds | 1 | 2 | 3 | 4 | 5 | 6 | 7 | 8 | 9 | 10 | Average |
|---|---|---|---|---|---|---|---|---|---|---|---|
| FPP | 26.55 | 44.40 | 47.13 | 43.17 | 43.81 | 49.39 | 52.55 | 43.71 | 42.94 | 44.54 | 43.82 |

**Table 16.4** Tenfold False Positives and the average for https traffic - Learning Phase II

| Tenfolds | 1 | 2 | 3 | 4 | 5 | 6 | 7 | 8 | 9 | 10 | Average |
|---|---|---|---|---|---|---|---|---|---|---|---|
| FPP | 57.26 | 60.07 | 59.95 | 57.58 | 57.18 | 56.94 | 54.89 | 55.34 | 53.82 | 55.70 | 56.87 |

mechanism helps us achieve suitable values for the geometric parameters in order to classify the traffic data efficiently. Within each learning phase, a cross validation technique is also applied on the labeled dataset and these phases are explained below.

*Cross Validation.* A tenfold cross validation is performed on the 80 % dataset. We partition the dataset into ten disjoint sets of equal size, and each time one set is excluded from the dataset and classification is performed on the remaining sets. In each learning phase we calculate the false positives for both http and https traffic for the tenfolds and take its average. This average is considered as the intermediate false positive for that particular phase. Below we present the results for four phases which reflect meaningful reduction in false positives. The results of all the learning phases are collected, but only that of four learning phases are presented here to clearly show the distinctions in false positive percentages (FPP).

*Learning Phase I.* In the first learning phase, the distance metric values are selected as $d_1=0$ and $d_2=0$ and they are validated. Note that the values of $(c_1, c_2)$ are always closer to $(-0.089, 0.602)$, hence its learning is not necessary. The learning is done by calculating the FPP for both http and https traffic after applying the LSVM and tenfold cross validation. Tables 16.1 and 16.2 show the FPP of all the tenfolds, and then the averages are calculated, which are presented in the last column of the tables. We can see that, when the distance metric parameters are set to zero, the FPP for http and https is very high. We also observe that the FPP for https is higher compared to that of http.

*Learning Phase II.* In this learning phase, the values of $d_1= 0.5$ and $d_2=0.1$ are assigned to the distance metric and its validation is learned by calculating the FPP for both http and https after the application of LSVM as discussed earlier. Tables 16.3 and 16.4 show the results of the FPP of all the tenfolds and, as mentioned

**Table 16.5** Tenfold False Positives and the average for http traffic - Learning Phase III

| Tenfolds | 1 | 2 | 3 | 4 | 5 | 6 | 7 | 8 | 9 | 10 | Average |
|---|---|---|---|---|---|---|---|---|---|---|---|
| FPP | 0 | 0 | 0 | 0 | 0 | 0 | 0 | 0 | 0 | 0 | 0 |

**Table 16.6** Tenfold False Positives and the average for https traffic - Learning Phase III

| Tenfolds | 1 | 2 | 3 | 4 | 5 | 6 | 7 | 8 | 9 | 10 | Average |
|---|---|---|---|---|---|---|---|---|---|---|---|
| FPP | 0 | 0 | 0 | 0 | 0 | 0 | 29.28 | 27.06 | 27.09 | 26.82 | 11.22 |

in the first phase, their averages are also calculated and presented in the last column of the tables. The results show significant decrease in the FPP values of the http traffic. However, FPP of https traffic is slightly increased. This is because the learning model is shifting only the rectangle of the http traffic to the left.

*Learning Phase III.* In this learning phase we assigned values for $d_1$ and $d_2$ as 1.01 and 0.1, respectively, and, as in the previous learning phases, the validation is learned and the average of FPP for http and https is calculated. Tables 16.5 and 16.6 show the false positives results for both http and https. In this learning phase the model was able to achieve no false positives for the http, however some false positives are observed for the https traffic. Therefore, the parameters $d_1$ and $d_2$ will be learned through more learning phases with cross validation technique.

*Learning Phase IV.* Finally, in this learning phase, with the proposed representation learning model we were able to achieve 0% false positives with LSVM classification for both the http and https traffic. In this case the learned values are $d_1 = 1.04$ and $d_2 = 0.1$ and they are validated by calculating the FPP after applying the LSVM. Therefore, the validation is complete and it is learned that the robust parameters for $d_1$ and $d_2$ are 1.04 and 0.1, respectively. Hence our final learned feature set for this application is $(-0.089, 0.602), 1.04, 0.1$.

In the learning phases we used a tenfold cross validation which in reality uses 90% of the 80% dataset. Hence, in the final phase of learning, we applied the learned parameters to the entire 80% dataset. The 80% dataset is shown in Fig. 16.5 and the corresponding isolated rectangles, using the learned parameters, are shown in Fig. 16.6. The LSVM classification is shown in Fig. 16.7 and it shows no false positives.

### 16.4.2   Testing Phase

In the training phase, we used the 80 % dataset and learned the robust values for the geometric parameters. In the testing phase, we use the remaining 20 % of the dataset to apply the techniques learned in the training phase. The results are shown below. Figure 16.8 shows the initial plot with the geometrical properties of the two classes.

Figure 16.9 illustrates the preprocessed datasets, where the parameters $d_1$ and $d_2$ are set to 1.04 and 0.1, respectively. Finally, Fig. 16.10 illustrates the result

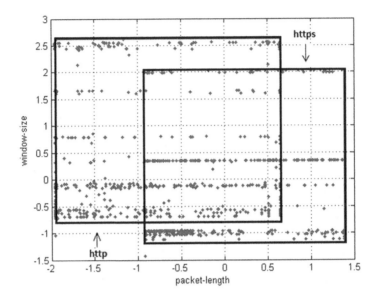

**Fig. 16.5** Geometrical properties of http and https traffic of 80% dataset

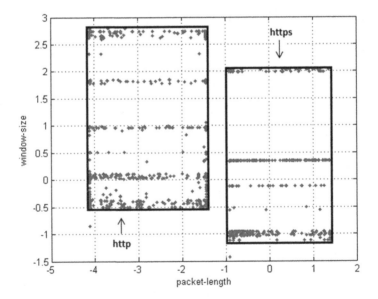

**Fig. 16.6** Isolated rectangles after preprocessing the 80% dataset

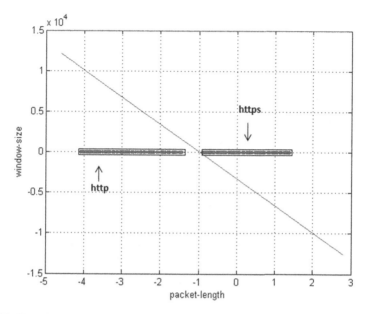

**Fig. 16.7** Classification of two traffic classes using LSVM on the 80 % dataset (rescaled Y-axis for the purpose of highlighting the SVM classifier)

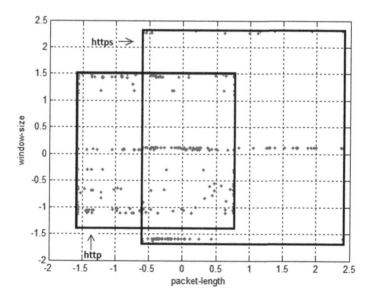

**Fig. 16.8** Geometrical properties of http and https traffic of 20% dataset

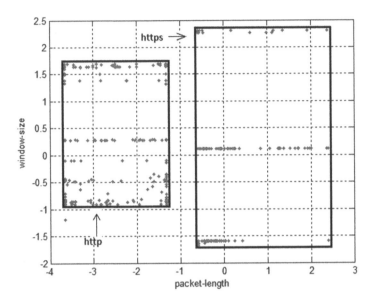

**Fig. 16.9** Isolated rectangles after preprocessing the 20% dataset

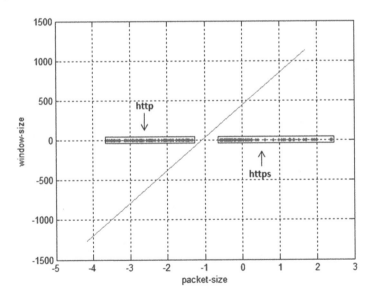

**Fig. 16.10** Classification of the two traffic classes using LSVM on the 20% dataset (rescaled Y-axis for the purpose of highlighting the SVM classifier)

of applying LSVM on the preprocessed dataset. The classification was done with high efficiency and no false positives were observed. This shows that the proposed approach is able to classify http and https traffic successfully.

## 16.5  Conclusion and Future Work

The primary goal of the study was to use the information readily available in the network traffic packets to classify TCP traffic and therefore reduce computational cost. The results show that the proposed approach was able to classify http and https traffic successfully based on information available in the TCP mechanism, i.e. packet-length and window-size. However, the dataset used for training and testing purposes may not be highly complicated and not large enough. Hence the learning model needs to be validated using more difficult and larger datasets. The study focused only on the classification of the http and https traffic and therefore it is useful to expand this learning model to other TCP protocols. Another area of future research is to use the other features available in the traffic flow and exploring the classification problems.

## References

1. Carela-Espaol, V., Barlet-Ros, P., Cabellos-Aparicio, A., Sol-Pareta, J.: Analysis of the impact of sampling on NetFlow traffic classification. Comput. Netw. 55(5), 1083–1099 (2011)
2. Corona, I., Giacinto, G., Roli, F.: Intrusion detection in computer systems using multiple classifier systems. Stud. Comput. Intell. 126, 91–113 (2008)
3. Cristianini, N., Shawe-Taylor, J.: An Introduction to Support Vector Machines Cambridge University Press, Cambridge, UK (2000)
4. Dainotti, A., Antonio P., Kimberly, C.C.: Issues and future directions in traffic classification. Netw. IEEE 26(1), 35–40 (2012)
5. Este, A., Gringoli, F., Salgarelli, L.: Support vector machines for TCP traffic classification. Comput. Netw. 53(14), 2476–2490 (2009)
6. Giacinto, G., Perdisci, R., Roli, F.: Network intrusion detection by combining one-class classifier. In: Roli, F., Vitulano, S. (eds.) ICIAP 2005. LNCS, Springer Verlag 3617, pp. 58–65 (2005)
7. Hastie, T., Tibshirani, R., Friedman, J.H.: The Elements of Statistical Learning: Data Mining, Inference, and Prediction. Springer, New York (2001)
8. Huang, G., Chen, H., Zhou, Z., Yin, F., Guo, K.: Two-class support vector data description. Pattern Recogn. 44, 320–329 (2011)
9. Laskov, P., Dussel, P., Schafer, C., Rieck, K.: Learning intrusion detection: supervised or unsupervised? In: Proceedings of the 13th ICIAP Conference, pp. 50–57 (2005)
10. LBNL/ICSI Enterprise Tracing Project: http://www.icir.org/enterprise-tracing/
11. Lee, S., Kim, H.C., Barman, D., Lee, S., Kim, C.K., Kwon, T.: Netramark: a network traffic classification benchmark. ACM SIGCOMM Comput. Commun. Rev. 41(1), 22–30 (2011)
12. Li, Y., Wang, J., Tian, Z., Lu, T., Young, C.: Building lightweight intrusion detection system using wrapper-based feature selection mechanisms. Comput. Secur. 28(6), 466–475 (2009)

13. Mangasarian, O.L., Musicant, D.R.: Lagrangian support vector machines. J. Mach. Learn. Res. **1**, 161177 (2001)
14. Quinlan, J.R.: C4.5: Programs for Machine Learning, Morgan Kaufmann, San Mateo, CA. (1993)
15. Tu, W., Sun, S.: Cross-domain representation-learning framework with combination of class-separate and domain-merge objectives. In: Proceedings of the CDKD12 Conference, pp. 18–25 (2012)
16. Zhang, H., Lu, G., Qassrawi, M.T., Zhang, Y., Yu, X.: Feature selection for optimizing traffic classification. Comput. Commun. **35**(12), 1457–1471 (2012)
17. Zhang, J., Xiang, Y., Wang, Y., Zhou, W., Guan, Y.: Network traffic classification using correlation information. IEEE Trans. Parall. Distr. **24**(1), 104–117 (2013)
18. Zhao, J.J., Huang, X.H., Sun, Q., Ma, Y.: Real-time feature selection in traffic classification. J. China U. Posts Telecommun. **15**, 68–72 (2008)
19. Zhen, L., Qiong, L.: A new feature selection method for internet traffic classification using ML. Phys. Procedia **33**, 1338–1345 (2012)
20. Zuev, D., Moore, A.: Traffic classification using a statistical approach. Passive Active Netw. Meas., **3431**, 321–324 (2005)

# About the Editors

**Dr. Maya Chhetri** is a professor of mathematics at the Department of Mathematics and Statistics at UNCG. She received her Ph.D. from Mississippi State University in 1999 and joined UNCG the same year. She works in the area of differential equations and nonlinear analysis. In particular, her research interest is in the study of positive solutions of nonlinear boundary value problems, both ODEs and elliptic PDEs and their applications to other disciplines.

**Dr. Sat Gupta** received Ph.D. in mathematics from University of Delhi (1977) and Ph.D. in statistics from Colorado State University (1987). He taught at University of Delhi for 6 years, at University of Southern Maine for 18 years, and has been at UNC Greensboro since 2004. He became a full professor in 1997. His main research area is sampling designs, particularly designs needed for collecting information on sensitive topics where there is a greater likelihood of respondent evasiveness and untruthfulness. He has collaborated with researchers from many fields including biology, marine biology, education, anthropology, psychology, medicine, nursing, and computer science. Some of these collaborative works have been funded by NSF, NIH, and other funding agencies. He is founding editor of the *Journal of Statistical Theory and Practice* (http://www.tandfonline.com/loi/UJSP20) besides serving on the editorial boards of several other journals.

**Dr. Jan Rychtář** is an associate professor of mathematics at the Department of Mathematics and Statistics at UNCG. He earned a Ph.D. in 2004 from the University of Alberta, and he joined the UNCG faculty the same year. He works in mathematical biology and game theory. With Mark Broom, he has coauthored a book "Game-theoretical models in biology" and authored or coauthored over 45 papers in peer-reviewed journals. Since 2005, he organizes an annual UNCG Regional Mathematics and Statistics Conference. He has supervised research of over 30 undergraduate students and has served as an Interim Director of the UNCG Office of Undergraduate Research in 2012–2013. He is a councilor for mathematics and computer sciences of the Council of Undergraduate Research.

J. Rychtář et al. (eds.), *Topics from the 8th Annual UNCG Regional Mathematics and Statistics Conference*, Springer Proceedings in Mathematics & Statistics 64, DOI 10.1007/978-1-4614-9332-7, © Springer Science+Business Media New York 2013

**Dr. Ratnasingham Shivaji** joined the University of North Carolina at Greensboro (UNCG) as H. Barton Excellence Professor and Head of the Department of Mathematics and Statistics in July 2011. Prior to joining UNCG, he served for 26 years at Mississippi State University (MSU), where he was honored as a W.L. Giles Distinguished Professor. He received his Ph.D. in Mathematics from Heriot-Watt University in Edinburgh, Scotland in 1981, and his B.S. (first class honors) from the University of Sri Lanka in 1977. Shivaji's area of specialization is partial differential equations and, in particular, nonlinear elliptic boundary value problems. His research work has applications in combustion theory, chemical reactor theory, and population dynamics and has been funded by the National Science Foundation. To date, Shivaji has authored or coauthored 117 research papers and served as thesis advisor for ten Ph.D. graduates.

Printed in the United States
By Bookmasters